# Engineering Education Quality Assurance

Arun S. Patil • Peter J. Gray
Editors

# Engineering Education Quality Assurance

## A Global Perspective

Springer

*Editors*
Arun S. Patil
Faculty of Sciences
Engineering and Health
CQUniversity Australia
Mackay Campus
12G.03, Boundary Road
Mackay QLD 4741
Australia
a.patil@cqu.edu.au

Peter J. Gray
Faculty Enhancement Center
589 McNair Road
10M, United States Naval Academy
Annapolis, MD 21402
USA
pgray@usna.edu

ISBN 978-1-4419-0554-3 e-ISBN 978-1-4419-0555-0
DOI 10.1007/978-1-4419-0555-0
Springer Dordrecht Heidelberg London New York

Library of Congress Control Number: 2009927500



Printed on acid-free paper

Springer is part of Springer Science+Business Media (www.springer.com)

# Preface

With the rapid globalization of higher education as well as related changes in social, political, economic, and other conditions over the last 25 years there have been ever increasing expectations for higher education, in general, and Engineering Education, in particular. These expectations are often expressed in terms of the need for Quality Assurance locally, regionally, and globally.

In some cases, there is a long tradition of independence and self-regulation of higher education institutions and programs. In other contexts, there has been considerable governmental regulation and disciplinary direction over time. The authors in this volume represent essentially all continents and 15 different countries. The common issues that they raise and their accounts of past, present, and future challenges provide a *snapshot* of the current state of Quality Assurance in higher education and Engineering Education.

This volume begins with an overview of the history and background of Quality Assurance in higher education and Engineering Education over the last century. The discussion of the historical, philosophical, political, and social background of Quality Assurance sets the stage for the other chapters. Following this broad brush stoke introduction, in the next part of the book, authors describe the general issues and challenges facing Quality Assurance in the twenty-first century from both regional and national perspectives. These authors have extensive experience in the area of Quality Assurance and have observed its growth and develop first hand over many years.

Next is a set of ten chapters that focus on individual countries. These chapters are written by leaders in Quality Assurance who know well the issues and challenges faced by their countries as they strive to meet both internal and external demands for Quality Assurance. It is clear from these chapters that there is much in common regarding the current state of Quality Assurance around the world.

In the last part of the book, a variety of strategies and techniques are described that can help develop and implement effective Quality Assurance approaches. The volume closes with a discussion of a conceptual framework for organizing internal and external Quality Assurance approaches for improvement and accountability. This chapter and the other chapters in the last part of the book are intended to provide Engineering Educators with a broad view of the tools and techniques available to meet a variety of expectations regarding Quality Assurance.

We would like to acknowledge the thought and effort that the contributing authors have made in drafting their respective chapters. Their good will in accepting our invitation to contribute to this volume and then their graciousness in responding to editorial suggestions and making revisions in a timely fashion is greatly appreciated. There are few volumes that bring together such an august and competence set of contributors. It is our hope that the insights into Quality Assurance in higher education and Engineering Education that our authors have given us as editors will be equally appreciated by our readers.

The support of the Springer editorial staff is also greatly appreciated. We could not have produced this volume without their expert guidance and technical assistance.

Mackay, Australia — Arun S. Patil
Annapolis, MD, USA — Peter J. Gray

# Contents

# Contributors

**Giuliano Augusti** ENAEE - European Network for Accreditation of Engineering Education and Università La Sapienza – Facoltà di Ingegneria, Via Eudossiana 18, I-00184 Roma, Italy, giuliano.augusti@uniroma1.it

**Naziha Ahmad Azli** Department of Energy Conversion, Faculty of Electrical Engineering, Universiti Teknologi Malaysia, 81310 UTM Skudai, Johor Malaysia, naziha@fke.utm.my

**Oleg V. Boev** Accreditation Center, Russian Association for Engineering Education, Office 328, 30 Lenin Avenue, Tomsk 634050, Russia, ovb@ac-raee.ru, ovb@tpu.ru

**Larry A. Braskamp** Loyola University Chicago, 820 N Michigan Avenue, Chicago, IL 60611, USA, larrybraskamp@yahoo.com

**Doris R. Brodeur** Department of Aeronautics and Astronautics, Massachusetts Institute of Technology, 77 Massachusetts Avenue, Cambridge, MA 02139, USA, dbrodeur@mit.edu

**Rosario Carrasco** Facultad de Ciencias Físicas y Matemáticas, Universidad de Chile, Beaucheff 850, Santiago, Chile, rosario.carrasco@usach.cl

**Alexander I. Chuchalin** Accreditation Center, Russian Association for Engineering Education, Office 328, 30 Lenin Avenue, Tomsk 634050, Russia, chai@tpu.ru

**Gary Codner** Faculty of Engineering, Monash University, Building 72, Faculty Office Wellington Road, Clayton, Melbourne VIC 3800, Australia, Gary.Codner@eng.monash.edu.au

**John Cowan** Academic Development, Edinburgh Napier University, Bevan Villa, Craighouse Campus, Edinburgh EH10 5LG, Scotland, J.Cowan@napier.ac.uk

**Edward F. Crawley** Department of Aeronautics and Astronautics, Massachusetts Institute of Technology, 77 Massachusetts Avenue, Cambridge, MA 02139, USA, crawley@mit.edu

**Salwani Mohd. Daud** Department of Energy Conversion, Faculty of Electrical Engineering, Universiti Teknologi Malaysia, 81310 UTM Skudai, Johor Malaysia, salwani@ic.utm.my

**Chrysanthe Demetry** Academic Affairs, Worcester Polytechnic Institute, 100 Institute Road, Worcester, MA 01609, USA, cdemetry@wpi.edu

**Henk Eijkman** The University of New South Wales at The Australian Defence Force Academy, Bldg 28 Rm 204, Canberra ACT 2600, Australia, H.Eijkman@adfa.edu.au

**Peter J. Gray** Faculty Enhancement Center, United States Naval Academy, 589 McNair Road, 10M, Annapolis, MD 21402, USA, pgray@usna.edu

**Hu Hanrahan** School of Electrical and Information Engineering, University of the Witwatersrand, Johannesburg, Private Bag 3, WITS 2050, South Africa, h.hanrahan@ee.wits.ac.za

**Kalayanee Jitgarun** Electrical Technology Education Department, Faculty of Industrial Education and Technology, King Mongkut's University of Technology Thonburi, Bangkok, Thailand, kalayanee.jit@kmutt.ac.th

**Obada Kayali** The University of New South Wales at The Australian Defence Force Academy, Bldg 28 Rm 204, Canberra ACT 2600, Australia, O.Kayali@adfa.edu.au

**Paiboon Kiattikomol** Electrical Technology Education Department, Faculty of Industrial Education and Technology, King Mongkut's University of Technology Thonburi, Bangkok, Thailand, ipaiomol@kmutt.ac.th

**Anastasia A. Kriushova** Accreditation Center, Russian Association for Engineering Education, Office 328, 30 Lenin Avenue, Tomsk 634050, Russia, kaa@tpu.ru

**Hao V. Le** Department of Educational Quality Assurance & Testing, Nha Trang University, 2 Nguyen Dinh Chieu, Nha Trang, Vietnam, haolevan@yahoo.com

**Mario F. Letelier** Centre for Research in Creativity and Higher Education, Universidad de Santiago de Chile, Casilla 10233, Santiago, Chile, mario.letelier@usach.cl

**Nik Hasnaa Nik Mahmood** Department of Energy Conversion, Faculty of Electrical Engineering, Universiti Teknologi Malaysia, 81310 UTM Skudai, Johor Malaysia, nikhasna@ic.utm.my

**Johan L. Malmqvist** Department of Product and Production Development, Hörsalsvägen 7A, Chalmers University of Technology, SE-412 96 Göteborg, Sweden, johan.malmqvist@chalmers.se

**Egon Moesby** School of Basic Studies, The Faculties of Engineering, Science and Medicine, Aalborg University, Strandvejen 12-14, DK-9000 Aalborg, Denmark, hem@aaue.dk

**Chenicheri Sid Nair** Centre for Higher Education Quality, Monash University, Wellington Road, Melbourne VIC 3800, Australia, Sid.Nair@adm.monash.edu.au

**R. Natarajan** 52/1, 13th Cross Road, Malleswaram, Bangalore 560 003, India, prof.rnatarajan@gmail.com

**Kim D. Nguyen** Institute for Educational Research, Ho Chi Minh City – University of Pedagogy, Ho Chi Minh City, Vietnam, nguyenkimdung@yahoo.com

**John A. Orr** Academic Affairs, Worcester Polytechnic Institute, 100 Institute Road, Worcester, MA 01609, USA, orr@wpi.edu

**Arun Patil** Faculty of Sciences, Engineering and Health, CQUniversity Australia, Mackay Campus, 12 G.03, Boundary Road, Mackay QLD 4741, Australia, a.patil@cqu.edu.au

**George D. Peterson** ABET, Inc., 111 Market Pl., Suite 1050, Baltimore, MD 21202, USA, gpeterson@abet.org

**Patricio V. Poblete** Facultad de Ciencias Físicas y Matemáticas, Universidad de Chile, Beaucheff 850, Santiago, Chile, ppoblete@ing.uchile.cl

**Marlia Puteh** College of Science and Technology, Universiti Teknologi Malaysia International Campus Jalan Semarak, 54100 Kuala Lumpur, Malaysia, marlia@citycampus.utm.my, marlia.puteh@gmail.com

**Palle Qvist** Department of Development and Planning, Aalborg University, Fibigerstræde 13, DK-9220 Aalborg, Denmark, pal@tnb.aau.dk

**Aija Sadurskis** Swedish National Agency for Higher Education, Högskoleverket Box 7851, SE-103 99 Stockholm, Sweden, aija.sadurskis@hsv.se

**Lance Schachterle** Academic Affairs, Worcester Polytechnic Institute, 100 Institute Road, Worcester, MA 01609, USA, les@wpi.edu

**Kin Wai Michael Siu** School of Design, The Hong Kong Polytechnic University, Hunghom, Kowloon, Hong Kong, m.siu@polyu.edu.hk

**Anuvat Tongsakul** Electrical Technology Education Department, Faculty of Industrial Education and Technology, King Mongkut's University of Technology Thonburi, Bangkok, Thailand, a_tongsakul@yahoo.com

**Algirdas Vaclovas Valiulis** Mechanical Faculty, Vilnius Gediminas Technical University, J. Basanaviciaus Street 28, LT-03224, Vilnius, Lithuania, algirdas.valiulis@adm.vgtu.lt

**Donatas Valiulis** Mechanical Faculty, Vilnius Gediminas Technical University, J. Basanaviciaus Street 28, LT-03224, Vilnius, Lithuania, donates.valiulis@adm.vgtu.lt

**Ximena Vargas** Facultad de Ciencias Físicas y Matemáticas, Universidad de Chile, Beaucheff 850, Santiago, Chile, xvargas@ing.uchile.cl

**Laurie Woollacott** School of Chemical and Metallurical Engineering, Faculty of Engineering and Built Environment, University of the Witwatersrand, Johannesburg, Private Bag 3, WITS 2050, South Africa, lorenzo.woollacott@wits.ac.za

**Stephen Yeomans** The University of New South Wales at The Australian Defence Force Academy, Bldg 20, Rm 135, Canberra ACT 2600, Australia, S.Yeomans@adfa.edu.au

# Overview/History

# The Background of Quality Assurance in Higher Education and Engineering Education

**Peter J. Gray, Arun Patil, and Gary Codner**

**Abstract** This chapter presents a review of the historical, philosophical, political, and social background of Quality Assurance of higher education, in general, and engineering education, in particular. Such a review can help us appreciate how the Quality Assurance movement got to where it is today and the tensions that are inherent in it, as well as provide guidance for its future development. Suggestions for advancing Quality Assurance in Engineering Education are provided at the end of the chapter.

## Introduction and Definition of Terms

At the *UNESCO World Conference on Higher Education in the Twenty-first Century: Vision and Action*, "Quality Assurance, accreditation, and the recognition of qualifications were identified as fundamental concerns for higher education" (López-Segrera 2007, p. xlvi). Evidence that Quality Assurance and accreditation are growing into worldwide, higher education phenomena include the formation of the International Network for Quality Assurance Agencies in Higher Education (INQAAHE) (Woodhouse 2007), the creation of the INQAAHE Guidelines of Good Practice in Quality Assurance, and the planned offering of an INQAAHE developed Graduate Certificate in Quality Assurance by the University of Melbourne. In addition, the Quality Assurance movement is clearly spreading to engineering education worldwide with the adoption of the Washington Accords 1989, Sydney Accords 2001, and Dublin Accords 2002 (International Engineering Alliance 2007).

And, it is clear from the range of countries, organizations, institutions, and authors represented in this book as well as the wealth of other recent publications; the vast variety of resources on sites such as Internet Rescurces for Higher

P.J. Gray (✉)
Faculty Enhancement Center, United States Naval Academy,
589 McNair Road, 10M, Annapolis, MD 21402, USA
e-mail: pgray@usna.edu

A.S. Patil and P.J. Gray (eds.), *Engineering Education Quality Assurance: A Global Perspective*,
DOI 10.1007/978-1-4419-0555-0_1, © Springer Science+Business Media, LLC 2009

Education Outcomes Assessment (North Carolina State University 2009); and even Google searches (12/26/08 at 1600 BST: 2,130,000 for Quality Assurance higher education in 0.28 s), that for better or worse *Quality Assurance*, *accreditation*, and the *recognition of qualifications* are truly the defining issues for higher education in the twenty-first century.

It is important to know how higher education Quality Assurance (QA), in general, and QA in engineering education, in particular, got to this point so that we can understand current conditions and thoughtfully guide the way forward. Quality Assurance encompasses some very complex concepts with multiple implicit and explicit meanings. Its various manifestations have had and increasingly will have profound implications for higher education professionals globally, nationally, institutionally, and individually. The impact will also be felt by various other higher education *stakeholders* including current and prospective students, parents and the general public, employers, and governmental and other Quality Assurance agencies including legislatures responsible for funding and overseeing higher education.

Definitions related to Quality Assurance that provide the context not only for the rest of this chapter but also for the other chapters in this book are discussed next. Then the various historical, philosophical, political, and social underpinnings of the Quality Assurance movement are the focus of the following section. Suggestions for advancing Quality Assurance in Engineering Education are the focus of the last section of the chapter.

## Definitions

To set the stage, some basic definitions of the key terms and concepts related to the Quality Assurance movement in higher education are discussed next. Quality Assurance has been defined most broadly by Harman and Meek (2000, p. 4) as:

> …systematic management and assessment procedures adopted by a higher education institution or system to monitor performance and to ensure achievement of quality outputs or improved quality.

This definition does not necessarily link assessment either formally or informally with accreditation, but other definitions define *accreditation* as a process of evaluating an institution or program to determine whether it meets accrediting body standards and if so granting recognition in the form of accreditation.

Similarly, Schwarz and Westerheijden (2007) define a Quality Assurance *scheme* or Quality Assurance *system* as "accreditation and evaluation systems *together*" (emphasis added, p. 3) by defining *accreditation* as (p. 2):

> institutionalized and systematically implemented evaluation schemes that end in a formal summary judgement that leads to formal approval processes regarding the respective institution, degree type and/or programme.

*Accreditation* is the element of Quality Assurance schemes that sets the standards for granting (Schwarz and Westerheijden 2007, p. 2):

> the "right to exist" within the system (or, respectively, to reject the "right to exist") to an institution, degree-type, programme (e.g., charter, license, accreditation).

In turn, *evaluation* activities are defined as (p. 3):

> institutionalized and systematically implemented activities regarding the measurement, analysis and/or development of quality for institutions, degree-types and/or programmes.

The terms *assessment* and *evaluation* are often used synonymously denoting both *means*, i.e., techniques, procedures, instruments, and methods for measurement and analysis used to monitor performance and, *ends*, "to ensure achievement of quality outputs or improved quality" (Harman and Meek 2000, p. 4).[1]

*Accountability* is another term that has been associated with such a definition and denotes a *responsibility or answerability to external audiences.*

The linking of Accreditation, Evaluation or Assessment, and Accountability in higher education Quality Assurance (QA) schemes causes considerable tensions because of their historical, philosophical, political, and social background. Understanding this background can help us, first, appreciate how the Quality Assurance movement got to where it is today and, second, how to guide its development in the future, given the tensions just noted.

## Background of Quality Assurance

"Quality in the sense of achieving academic excellence has always been a central value in higher education" (Schwarz and Westerheijden 2007, p. 4). Institutions of higher education have their beginning relied on the reputation of their faculties to attract students and scholars and to give credibility to their degree programs, their graduates, and their researches.

However, the way Quality Assurance's key components, Accreditation and Evaluation or Assessment, are defined has a great influence on its implementation and impact. Assessment is about *language* regarding the nature of teaching, learning, and appropriate inquiry and *power* regarding how higher education is organized and rewarded (Ewell 1989). Quality Assurance, Accreditation, and Accountability are all implied in Ewell's definition of Assessment (see footnote 1).

Four very broad traditions in higher education comprise the key strands of the historical, philosophical, political, and social foundations of Quality Assurance. The first is academic peer-review-based Accreditation, the second is governmental

[1]The terms evaluation and assessment (lower-case *e* and *a*) can refer to a set of techniques, procedures, instruments, and methods for measurement and analysis. These are used in formal Evaluation, Assessment, Accreditation, and Quality Assurance schemes (upper-case *E*, *A*, and *QA*) *to monitor performance* and *to ensure achievement of quality outputs or improved quality*. In this sense, Evaluation or Assessment are synonymous with Quality Assurance as proper nouns denoting a movement, process, approach, or even a profession (such as is embodied in the American Evaluation Associate professional standards or International Network Quality Assurance Agencies in Higher Education certificate program).

oversight, the third includes the Scientific Education and Management Movements, and the fourth is the Accountability movement. Unless these different traditions and their related language and power implications are clearly understood and addressed, it is likely that conflicts will arise that could severely inhibit the potential positive impact of Engineering Education Quality Assurance as it spreads around the world.

## *The Foundation of Accreditation: Professional Authority*

Quality Assurance of US higher education, based on a scheme of professional authority gained through experience, began in the late 1800s. The North Central Association of Schools and Colleges was the first voluntary accrediting association.[2] Therefore, instead of a nation-wide governmental system of higher education Quality Assurance, regional associations were established in the USA that reflect the cultures of their constituent members.

Similarly, QA in engineering and technology programs began in the USA as a voluntary effort organized by the Accreditation Board for Engineering and Technology (ABET, Inc.) in 1936.[3] Historical evidence of engineering accreditation in Europe is the Law of 10 July 1934 implemented by *La Commission des Titres de l'Ingénieur* (Commission of the Titles of Engineer) in France related to the conditions of delivery and the use of the title of qualified engineer (CTI 2006).

Accreditation and Quality Assurance processes in Europe have their roots in the 1950s, when several initiatives at regional and national levels were carried out in the form of educational audits intended to assess pedagogical skills in higher education (Irandoust et al. 2000). The establishment of the European Federation of National Engineering Associations (FEANI) in 1951 was an important initiative intended to foster a common accreditation approach in Europe. However, as noted by Augusti (2007, p. 101), "The word accreditation, used in the USA since the 1930s, did not find its way into European specialized literature and official documents until very recently, but since then it has rapidly become a catchword." The same is true for other regions of the world.

In the chapter "EUR-ACE: the European Accreditation system of Engineering Education and its Global Context," Augusti explains that the European Commission first acknowledged the possible synergies between the recognition of qualifications for academic and professional purposes in 1994. Then in 1998–1999 the Thematic

---

[2]See the CHEA Web site for an overview of US accreditation, http://www.chea.org/pdf/overview_US_accred_8-03.pdf and for a directory of CHEA Recognized Organizations http://www.chea.org/Directories/index.asp

[3]See the chapter "Quality Assurance in the Preparation of Technical Professionals: The ABET Perspective" by Peterson for a discussion of ABET, Inc. and the chapter "Quality Assurance in Engineering Education in the United States" by Schachterle for an overview of higher education and engineering education accreditation in USA.

Network, Higher Engineering Education for Europe (H3E) organized three European Workshops for Accreditation of Engineering Programs which in turn lead to the establishment in September 2000 of the European Standing Observatory for the Engineering Profession and Education (ESOEPE). The definition of accreditation adopted by FEANI in 2001, and largely accepted by the engineering education community in Europe, outlines the relationship between Accreditation and Quality Assurance (ESOEPE 2005):

> Accreditation is the primary Quality Assurance process used to ensure the suitability of an educational programme as the entry route to the engineering profession. Accreditation involves a periodic audit against published standards of the engineering education provided by a particular course or programme. It is essentially a peer review process, undertaken by appropriately trained and independent panels comprising both engineering teachers and engineers from industry. The process normally involves both scrutiny of data and a structured visit to the educational institution.

Still, within Europe there were great differences in the existing accreditation procedures that have led to confusion and difficulties in the mobility of engineering professionals. It was not until 2006 that a framework for establishing a European system for accreditation of engineering education was completed as part of the EUR-ACE (EURopean ACredited Engineer) project.[4] While its purpose is generally similar to other accreditation schemes, the EUR-ACE framework is specifically related to the first and second cycles (Bachelors and Masters degrees) as defined within the Bologna process and has the specific aims of (Augusti 2007, p. 101):

- Providing an appropriate "European label" to accredited educational programs
- Improving the quality of educational programs in engineering
- Facilitating transnational recognition by the label marking
- Facilitating recognition by the competent authorities, in accord with EU directives
- Facilitating mutual recognition agreements

Within the larger Bologna process, the standards and guidelines for Quality Assurance in higher education have been developed by the European Association for Quality Assurance in Higher Education (ENQA 2007). However, there is still considerable variation in accreditation standards and practices across Europe and Russia.[5]

The major concerns regarding accreditation in Asia-Pacific higher education systems are inconsistency from country to country, lack of mutual recognition, and

[4] Augusti describes the history, current status, and future development of EUR-ACE in the chapter "EUR-ACE: The European Accreditation system of Engineering Education and its Global Context."

[5] See the general discussions by Augusti ("EUR-ACE: the European Accreditation system of Engineering Education and its Global Context") and Cowan ("Quality Assurance in European Engineering Education: Present and Future Challenges") and the specific descriptions for Sweden by Malmqvist and Sadurskis ("Quality Assurance of Engineering Education in Sweden"), Lithuania by Valiulis and Valiulis ("Engineering Education Quality Assurance: The Essential Pillar of Higher Education Reform in Lithuania"), and Russia by Chuchalin et al. ("Quality Assurance in Engineering Education and Modernization of Higher Education in Russia").

slow rate of development and implementation. While countries like Australia and New Zealand have quite well-developed QA systems, only recently other countries in this part of the world have taken steps to establish QA schemes comparable to those just described. For example, within the last few years the Japan Accreditation Board of Engineering Education (JABEE), the Accreditation Board for Engineering Education of Korea (ABEEK), Institute of Engineering Education Taiwan (IEET), and the Institute of Engineers, Singapore joined the Washington Accord group.

Other Asia-Pacific countries as well as those in Latin America[6] are still evolving from a system of governmental controls as the primary means of Quality Assurance.

## *Nature of Accreditation*

When Accreditation as a formal process was begun during the early part of the twentieth century, the *language* was that of self-regulation by higher education faculty members themselves to assure quality and foster improvement. That is, Accreditation was a peer-review process based on professional authority gained through experience and, as such, the *power* was in the hands of professional educators (Gray 2002). This is a "subjectivist and intuitivist ethic that values the tacit knowledge of professional authorities" (Gray 2002, p. 51).

The heart of the Accreditation process is the institutional or program self-study (i.e., self-assessment or evaluation) and site visit by peers. The intent of the site visit is to *monitor performance* through an *evaluation* by external colleagues (peers) related to the observations of the self-study undertaken by internal stakeholders, as well as first-hand experience during a site visit. Recommendations for improvement are provided to internal audiences related to the Accreditation standards and a recommendation about accreditation status is made to the accrediting agency.

Accreditation agencies, for example, the US regional institutional accrediting agencies and discipline-specific associations, such as ABET, Inc., are guided by boards of directors comprised of representatives of relevant institutions or disciplines and are essentially *membership organizations*. Quality standards set by accrediting agency boards guide both the self-study and site visit that lead to a decision about accreditation, i.e., the right to exist within the system. This tradition is especially

[6]See specific discussions on India by Natarajan ("Assessment of Engineering Education Quality: An Indian Perspective"), Vietnam by Le and Nguyen ("Quality Assurance in Vietnam's Engineering Education"), Malaysia by Puteh et al. ("Quality Issues Facing Malaysian Higher Learning Institutions: A Case Study of Universiti Teknologi Malaysia"), Thailand by Jitgarun et al. ("Quality Assurance for the Engineering Para-Professional in Thailand"), Hong Kong by Siu ("Quality Assurance in Engineering Education: An All-round Perspective"), and Chile and Latin America by Letelier et al. ("Quality Assurance in Higher Education in Chile: National and Engineering Dimensions").

strong in the USA with its history of autonomy and diversity of higher education institutions.

In institutional and disciplinary Quality Assurance schemes that include Accreditation, specific recommendations intended to guide the *development or improvement of quality* are not typically disclosed to anyone outside the institution. If any information is shared with external audiences, it is primarily a quite general categorization of accreditation status, for example, accredited, accredited with reservations, and not accredited.

This self-regulation by peer review was a successful and trusted means of assuring higher education quality through the 1970s.

## Governmental Oversight

In Europe and elsewhere, the control of quality in higher education has traditionally been through bureaucratic means based on government-provided budgets. As a result, there were "legal conditions for the establishment of institutions, faculties, and/or programs of study and state-provided means (funding, housing) to fulfill those conditions" (Schwarz and Westerheijden 2007, p. 4).

As a governmental function, the support of education is justified as a *public good*, i.e., an enterprise that benefits society in general. This is based on the assumption that the cost of providing education to its citizens is an investment that a country makes which pays dividends, implicitly, through an informed citizenry and, explicitly, though a country's advancement and increased wealth. In other words, "taxes are the price that we pay for goods and services produced in the public sector from which we all benefit" (Brooks 2005, p. A15)

In state-sponsored systems the *right to exist* is granted by a governmental agency or government authorized agency based on a set of, sometimes, legislated criteria and standards. An institution, degree type, and/or program must initially meet these expectations in order to operate and then must regularly submit reports documenting its *quality* in relation to set criteria and standards, in order to continue to operate.

In the USA, there are really two governmental systems, one at the Federal level and another at the individual state level. The Federal system has traditionally been conducted by external agencies authorized by the Federal government to conduct peer-review-based accreditation (see footnote 2). Then there is the oversight of education conducted by the 50 state governments. Education is a states' rights issue addressed in the US Constitution and Bill of Rights (see the chapter "Quality Assurance in Engineering Education in the United States" by Schachterle et al.).

The state systems of higher education began with the establishment of land-grant institutions in the 1860s and expanded dramatically in the 1950s, 1960s and 1970s for returning service men and women and their baby-boom offspring. These systems mirror the nation-wide governmental systems of higher education elsewhere in the world, in that, there are education agencies in each of the 50 states that grant the right to exist and periodically review programs. They also provide salaries for teachers as well as other funding such as capital construction.[7]

Until recently, the governmental oversight has not been overly intrusive. Nevertheless, having two sources of power related to Accreditation has caused tensions that foreshadow the kinds of issues that will have to be faced as the Quality Assurance movement goes global. In particular, these tensions concern how to reconcile professional peer review (internal, improvement-focused) and bureaucratic governmental (external, accountability-focused) approaches. These tensions have been heightened in the last 20 years as a result of the introduction of the Scientific Education and Management Movements and, ultimately, the accountability movement into higher education.

## *Scientific Education and Management Movements*

During the twentieth century, at the same time that the Accreditation movement was evolving (with its subjectivist and intuitivist ethic), another philosophical tradition developed, based on objectivist and utilitarian assumptions, that fostered the Scientific Education and Management Movements.

### Scientific Education Movement: Early Twentieth Century

Beginning with Ralph Tyler in the early twentieth century, the Scientific Education Movement used the same language as the professional authority-based Accreditation movement, but with a different power arrangement. Tyler described scientific education as the use of educational outcomes in the form of student behaviors, "to serve as the objectives for teaching and as the basis for testing" (Merwin in Gray 2002, p. 12). This is a rational empiricist process where it is assumed that educational outcomes are knowable in advance, specific, measurable, and related to behaviors that can be directly observed.

From the beginning, the scientific movement in education has had the following purposes (Tyler in Merwin 1969, p. 11):

- To monitor
- To help select or differentiate among

[7]For example, The New York State Office of College and University Evaluation (OCUE), oversees all degree-granting colleges and universities in New York State, and assures that the programs they offer for credit meet or exceed minimum quality standards. The Office's computerized database contains information on nearly 25,000 separate college programs. The Board of Regents Authority for Quality Assurance in Higher Education is based on various state laws and Regents Rules. See the New York State Education Department Office of Higher Education Web sites: http://www.highered.nysed.gov/ocue/ and http://www.highered.nysed.gov/ocue/board_of_regents_authority_for_q.htm

- To assist in development or improvement
- To identify the differential effects on different populations
- To provide estimates of effects and costs to consumers
- To test relevance and validity of principles upon which "programs" are based
- To facilitate discussion about innovation and change

The tensions introduced by the Scientific Education Movement into higher education are the foundation of Ewell's comments about the *language* and *power* aspects of Assessment. That is, while this list of purposes is similar in *language* to that used in the peer-review approach to accreditation and its related self-evaluation or assessment, there are very different *power* connotations. First, it goes beyond *monitoring* to comparing and contrasting educational institutions, programs, etc. And, it goes beyond *assisting in development or improvement* to the empirical measurement of the *effects* of education *on different populations* as well as estimates of *effects and costs* that are to be shared with consumers, presumably to help them decide which institutions, programs, etc. to choose.

This empirical approach also suggests that those with a scientific approach to education are the best able to determine its value and worth and that these determinations should be used as the basis of Quality Assurance. Such a perspective underpins the Accountability movement that will be discussed shortly. This mixed bag of purposes has had considerable impact over the years.

## Scientific Education Movement: 1960s and 1970s

During the 1960s and 1970s in USA, the Scientific Education Movement spawned objective testing and measurement methods, the use of behavioral objectives (Popham and Baker 1970), the establishment of organizations such as Educational Testing Services, and large-scale studies of educational impact (Worthen and Sanders 1973). In this context, many different evaluation models and theories were developed including the Context–Input–Process–Product (CIPP) model (Stufflebeam et al. 1971), the countenance of education evaluation or discrepancy evaluation model (Stake 1967, 1991), and the concepts of formative and summative evaluation (Scriven 1967). These all had considerable influence in the US and around the world. In fact, evaluation became more than a set of measurement and analysis methods; it became a movement with professional associations such as the American Evaluation Association, The Canadian Evaluation Society, and the African Evaluation Association, with advanced degree granting programs, and professional standards.[8]

In a 1975 paper, Stake foreshadowed all of the issues that we are currently facing with the Quality Assurance movement saying, "people expect evaluation to accomplish many different purposes" (1975, p. 7):

---

[8] See the Web site of the American Evaluation Association: http://www.eval.org/

- To document events
- To record student change
- To detect institutional vitality
- To place the blame for trouble
- To aid administrative decision making
- To facilitate corrective action
- To increase our understanding of teaching and learning

## Scientific Education Movement: 1980s and 1990s

In the 1980s, student learning outcomes assessment emerged under the leadership of the American Association for Higher Education (AAHE). As noted by Marchese (1987, p. 4), then vice-president of AAHE and editor of the *AAHE Bulletin*, "Assessment is not something just invented: a rich variety of approaches to knowing about student learning has evolved, through decades of research and campus experience" based on scientific education methods.

Palomba and Banta (1999) describe *Assessment* as *means*, the planned examination of information, and, *ends*, using this information to shape institutional policies, processes, and practices to help improve student and institutional performance. This is very much like current broad definitions of Quality Assurance. Ewell's use of Assessment should be understood within the context of the student learning outcomes assessment approach to Quality Assurance. Leaders such as Banta, Ewell, and the many thousands of higher education professionals from all over the world who attended the AAHE Assessment Forums in the 1980s and 1990s moved *Assessment* into the mainstream of higher education globally.[9]

Of course, instructors have always assessed students in relation to the content of their courses through a variety of means, formal and informal as well as qualitative and quantitative. And, like peer-review-based accreditation, setting the criteria and standards has traditionally been the prerogative of the faculty member based on tacit knowledge of the subject and intended learning related to a whole range of knowledge, skills, and dispositions of an educated person. Similarly, institutions have had curriculum committees or other internal structures for authorizing the establishment of particular programs of study or specific courses and a periodic review process to monitor quality and guide improvement. These structures, along with the peer review of scholarly and creative work and the peer-review accreditation process, based on professional authority gained through experience, are woven into the fabric and culture of higher education. It is not surprising then that the Assessment movement of the 1980s and 1990s clashed mightily with this traditional culture of higher education.

[9] See *Building a scholarship of assessment* by Banta and Associates (2002).

## Scientific Education Movement: 1990s to 2000s

Embedding a culture of Assessment in higher education has been and remains a major challenge. This challenge was taken up at the end of the twentieth century when the Assessment movement was adopted by many different Quality Assurance organizations. They began to include requirements for student learning outcomes assessment in their standards and practices. For example, ABET included a specific criterion on program learning outcomes and assessment in its 1998 revision of the criteria for accreditation called Engineering Criteria 2000 (*EC2000*). (Chapters "Quality Assurance in the Preparation of Technical Professionals: The ABET Perspective" by Peterson and "Quality Assurance in Engineering Education in the United States" by Schachterle describe in detail the learning outcomes criteria (a)–(k) and the impact of *EC2000* on Engineering Education in USA.)

In Europe, the Berlin Communiqué of 2003 directed the creation of "an over-arching framework of qualifications for the European Higher Education Area" to be described "in terms of workload, level, learning outcomes, competences, and profile" (Joint Quality Initiative 2004, p. 1) The complete set of *Dublin descriptors*, created in October 2004, included such qualifications (i.e., learning outcomes and competencies).

Between 1998 and 2001, the UK Quality Assurance Agency (QAA) Code of Practice[10] was prepared as "a statement of good practice that has been endorsed by the higher education community" (The Quality Assurance Agency for Higher Education 2006, p. 1). At the foundation of the Code is the assessment of student learning as described in Code Section 6:

> In higher education, "assessment" describes any processes that appraise an individual's knowledge, understanding, abilities or skills. There are many different forms of assessment, serving a variety of purposes. These include:

- Promoting student learning by providing the student with feedback, normally to help improve his/her performance
- Evaluating student knowledge, understanding abilities, or skills
- Providing a mark or grade that enables a student's performance to be established. The mark or grade may also be used to make progress decisions
- Enabling the public (including employers), and higher education providers, to know that an individual has attained an appropriate level of achievement that reflects the academic standards set by the awarding institution and agreed UK norms, including the frameworks for higher education qualifications. This may include demonstrating fitness to practice or meeting other professional requirements.

---

[10]Code of practice for the assurance of academic quality and standards in higher education (Code of practice) for the guidance of organizations subscribing to the Quality Assurance Agency for Higher Education (QAA) and other bodies offering UK higher education (The Quality Assurance Agency for Higher Education 2006).

At the same time as the concepts of the Scientific Education Movement were being adopted by accreditation agencies, another Quality Assurance scheme from business and industry emerged in higher education as is described next.

## Scientific Management Movement in Higher Education

Quality Assurance as a separate instrument in university management and government policy started in the 1970s and 1980s when it was discovered as a new management tool in industry mimicking the successes of the Japanese economy (Schwarz and Westerheijden 2007, p. 5). The result has been the adoption of such approaches as management by objectives (MBO) and total quality management (TQM) in higher education institutions.[11]

Similarly, an Input–Process–Output (IPO) framework that stems from consumer behavior theory has been applied to Quality Assurance in higher education (Chua 2004). The important difference between Chua's IPO framework and the Educational Process Cycle shown in Fig. 1 is the feedback loop from the Output part of the framework back to the input and teaching/learning process (Chua 2004; Patil and Codner 2007, 2008). This is sometimes called *closing the assessment loop.*

Such a cycle is based on the assumption that production in social services such as education is equivalent to production in business and industry (House in Gray 2002).

Hoecht (2006, p. 542) quotes from Habermasian's *The University in the New Corporate World* in which he:

> argues that the academic lifeworld, traditionally shaped by peer processes, academic freedom and the pursuit of knowledge, has been colonised by a (new) public sector managerialism.

As a result, the adoption of such objectivist and utilitarian approaches in higher education has caused great tensions that are well documented [Henkel and Chandler et al. cited in Hoecht (2006)].

In summary, the Scientific Education and Management Movements stem from objectivist and utilitarian assumptions. This is in contrast to the subjectivist and intuitivist assumptions of the Accreditation Movement that is based on professional authority gained through experience. Given the traditional culture of higher education, many faculty members, even if they are in scientific disciplines, hold subjectivist and intuitivist assumptions about how to organize and evaluate or assess teaching and learning and who should have the power to initiate such activities.

In addition, traditionally, Evaluation or Assessment for accreditation purposes examined the *capacity* of a higher education institution, degree type, or program to meet certain criteria and standards in relation to *inputs* and *processes*, i.e., the quality of resources and activities. However, with the introduction of the Scientific Education and Management Movements, and the adoption of student learning out-

---

[11] Hoecht (2006, p. 548) characterizes TQM in higher education as "a clash of principal assumptions and the difference between quality management for learning and quality management for control." This again brings up issues of language and power.

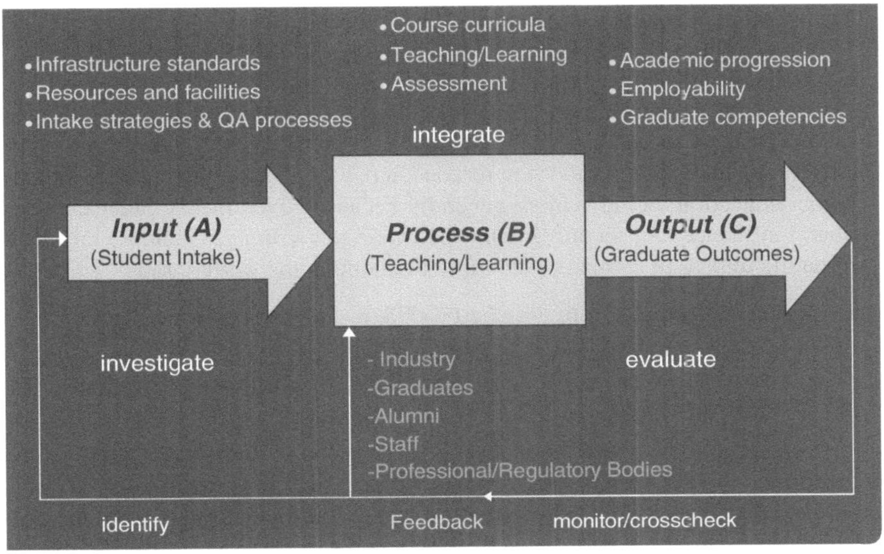


**Fig. 1** The Educational Process Cycle modified from IPO framework of educational quality

comes assessment as a means of Quality Assurance, the emphasis shifted to *outputs*. That is, the quality of graduates in terms of academic results (learning) and employability or workplace recruitments become the focus.

Such a change in the operational definition of *appropriate inquiry* and how higher education is judged and thus *rewarded* has caused additional apprehension. In addition, the stress caused by the introduction of the Scientific Education and Management Movements into higher education has been exacerbated by the rise of the Accountability movement described in the next section.

## *The Rise of the Accountability Movement*

The Scientific Education and Management Movements provided the philosophical context for the US *undergraduate reform* reports of 1984–1985 during the Reagan administration in the USA. Ewell notes that these reports made two assertions. The first is that individual student learning can be significantly enhanced through frequent communication about performance, *which is supported by research* and, second, that organizational change can occur, i.e., institutions can learn through information about results and can make continuous improvements in response, *which is not supported by research* (Ewell 1991).

And, around the same time, the first governmental policies related to Accountability were implemented in Western Europe. "Beside the usual reasons of copying whatever was started in the USA and, now, whatever was deemed successful in business, the following reasons underlie the adoption of these governance tools in Europe" (Schwarz and Westerheijden 2007, p. 5):

- Massification of higher education
- Limits of central control were reached with these larger higher education systems (that had developed after the Second World War)
- Deregulation was in fashion at the time when neo-liberalism (conservatism in the US) made a forceful entry into the political arena
- Governmental budget limits were reached, again because of the massification of higher education, but also more generally because governments under the neo-liberal influence (conservative in the US) were not willing to increase the share of public to private earnings even more to maintain the welfare state.[12]

These reasons are supported by Reichert (2008, p. 5) who points out that:

> Before Bologna, higher education debates in the 1990s were characterised by multiple national debates on quality problems in higher education, largely due to the effects of under-funded massification.

Underfunding and massification led to concerns about high student–staff ratios and resulting overcrowding of classrooms. These conditions, together with "outdated teaching methodologies and teacher-centered curricula, long study duration and high drop-out rates" led many to see higher education as not being able to respond to the demands of the times (Reichert 2008, p. 5). Massification was also the stimulus for the development of the taxonomy of engineering graduate's attributes and capabilities described in chapter "Taxonomies of Engineering Competencies and Quality Assurance in Engineering Education" by Woollacott.

Neal-Sturgess (2007, p. 129) adds, that as the background to the Bologna process,

> there was considerable concern in the 1990s at governmental level in the EU that Italy, Germany, France and many New Accession States have economically unsustainable, grossly inefficient higher education systems. Also, that the European higher education system was not making a sufficient contribution to the wealth creation process in the EU, and that the EU higher education system was hidebound and resistant to change.

In this regard, Reichert (2008, p. 5) concludes:

> At the same time more and more systems saw the need for increased autonomy of higher education institutions to enable them to face the widening range of demands and accelerating pace of international research competition better. The introduction of institutional autonomy and the simultaneous cutting back of state control could only be realised, however, in conjunction with heightened accountability provisions. Hence, in many countries Quality Assurance agencies were either created or transformed to meet these new demands.

This emphasis on *Accountability*, i.e., value for the money as measured by objective output data, has not only come with higher education institutions being given "autonomy to *do more with less*" (Schwarz and Westerheijden 2007, p. 5), but also more importantly such policies suggest a breakdown of the trust that society has traditionally had in the quality and value of higher education.

---

[12]Schwarz and Westerheijden (2007) quotations included with kind permission of Springer Science and Business Media. No part of this work may be reproduced, stored in a retrieval system, or transmitted in any form without written permission from the Publisher.

In this regard, the Accountability movement redefined higher education from being primarily a *public good* that deserved, if not required public financing for the advancement of society, to a *private good*, which is to the benefit and, therefore, is the responsibility of the individual. The rationale being that since further education results in increased earnings for individuals then they should bear the burden of its costs. Of course, as noted above, this calculation left out the contribution that higher education, and more generally the education of a country's citizens, makes to the wealth and advancement of society. There is some question as to whether this is a widely held view in society since higher education is an aspiration for an ever increasing number of people, which brings with it other responsibilities and challenges. And, within the context of the current global economic crisis, leaders all over the world have called for investments in higher education in order to stimulate recovery and prepare workers for the new economy that emerges.

In any case, over the last 25 years, this change in perspective had led to the decreased taxes and subsequently the reduction in funding of many programs for the public good. Affected were not only higher education, but also many previously supported public services including elementary (primary) and secondary education, health care, public transportation, infrastructure, and environmental protection.

Hoecht (2006) in examining the issues related to *auditing*, *accountability*, and *trust* concludes that while (p. 541):

> accountability and transparency are important principles that academics should wholeheartedly embrace… the audit format adopted in the UK introduces a one-way accountability and provides "rituals of verification" that instead of fostering trust, have high opportunity costs and may well be detrimental to innovative teaching and learning.

Another form of Accountability, Institutional Effectiveness Assessment came into existence in the 1980s and followed "wave after wave of imported business techniques such as MBO, PPBS, zero-based budgeting, and strategic planning" (Ewell in Gray 2002, p. 50) as noted above. Institutional Effectiveness Assessment rests on the foundation of student learning outcomes assessment in which it examines institutional policies, structures, and practices in light of the extent that they foster intended learning outcomes. Accountability is inherent in the Institutional Effectiveness Assessment movement since such schemes require institutions to report publicly (at least to the accrediting agencies, if not to external stakeholders) information about their effectiveness. Criteria for effectiveness may include graduation rates, time to degree, and employment of graduates as well as the extent of student achievement related to a set of internally or externally specified learning outcomes. This approach has even led to demands for standardized testing of college and university students in relation to such areas as general knowledge, critical thinking, and written communication.

In the USA, a Voluntary System of Accountability (VSA) has been developed through a partnership between the American Association of State Colleges and Universities (AASCU) and the National Association of State Universities and Land-Grant Colleges (NASULGC), from which most of the engineering professionals graduate (Voluntary System of Accountability 2009). The American Association of Colleges and Universities which represents liberal arts colleges and universities has for the past 25 years called on "the academy to take responsibility for assessing the quality of student learning in college" (AAC&U Board of Directors 2008, p. 1). However, in affirming that accountability is essential, AAC&U asserts that "the form it takes must be worthy of our mission" (p. 3). That is, it must be respectful of the learning outcomes that are essential to a liberal arts education as articulated in their ten recommendations for a new accountability framework (AAC&U Board of Directors 2008, pp. 13–14). In effect they are calling for *Assessment – of the Right Kind* (Lederman 2009).

At the same time, new rules proposed by the outgoing United States Secretary of Education within the area of accountability point to the possibility of drastic changes in the traditional institutional–federal relationship in the USA.

> The new law rearranges the institution–federal relationship in two major ways. Similar to what has happened with accreditation, institutions now have (1) a host of new areas of reporting and (2) expanded reporting in areas that are already in the law, culminating in 110 new reporting, record-keeping and regulatory requirements.

## Rankings and League Tables

An even more extreme manifestation of the Institutional Effectiveness Assessment and Accountability movements takes the form of rankings and league tables produced by organizations external to higher education such as *US New and World Report* (America's Best Colleges) and the *Times Higher Education* (THE) *Supplement* (World University Rankings). Usher and Savino (2007, p. 23), having reviewed 17 university league tables and ranking systems from around the world, note that (Eaton 2008):

> University rankings or "league tables," a novelty as recently as 15 years ago, are today a standard feature in most countries with large higher education systems. They were originally created over 20 years ago by *U.S. News & World Report* in order to meet a perceived market need for more transparent, comparative data about educational institutions.

However, these efforts have not always had the desired effect. Clarke (2007, p. 28) points out that in relation to one of the main avowed purposes of the rankings, "to remove economic, academic, and other barriers to access for particular student populations," they have, in fact, contributed to an "increasing stratification of the US higher education system by creating incentives for schools to recruit students who will be "assets" in terms of maintaining or enhancing their position in the rankings" (p. 38).

Cheng and Liu (2008) used *The Berlin Principles on Ranking of Higher Education Institutions*[13] to analyze 18 such efforts and provide 14 criteria for the development and use of rankings. They note that "While the ranking of higher education institutions (HEIs) has become more and more popular, there are

[13] An appendix to *College and University Ranking Systems* by Usher and Savino (2007) includes *The Berlin Principles on Ranking of Higher Education Institutions.*

increasing concerns about the quality of such ranking" (p. 201). However, after following the growth of rankings over the last decade, Sadlak et al. (2008, p. 195) conclude, "There is now increasing evidence that ranking systems are here to stay, and are having a growing effect on global dialogs about higher education quality and accountability."

A report by Professor David East, Chief Executive of the Higher Education Funding Council for England (HEFCE) on *Understanding Instructional Performance* states that (East 2008, p. 54):

(a) The use of performance measures should acknowledge that the costs and distortions tend to increase over time while the benefits diminish...
(b) Where it is possible to anticipate perverse incentives created by a new measure, "early warning" systems should be developed that will pick up distorted patterns of activity
(c) It is healthy for individual universities and colleges to take different approaches to performance measurement. HEFCE policy should encourage diverse management approaches to the problem of understanding performance.

Stufflebeam (in Gray 2002), drawing on his long experience with the Evaluation movement, said as much when he observed that the objectivist methods inherent in evaluation when used to hold courses, programs, or institutions accountable for learning can result in "invidious comparisons and thereby produce unhealthy competition and much political unrest and acrimony" (p. 20).

In summary, as suggested by the chronological layout of this section, there has been a steady shift toward external accountability over the last 20 years in higher education, in general, and engineering education, in particular. The rationale for this movement is public policy concern with the effectiveness of the funds invested, in part a function of rising costs and increasing complexity of higher education. In this context, Quality Assurance (QA) has come to be seen as a tool not just "to ensure achievement of quality outputs or improved quality" (Harman and Meek 2000, p. 4), but as a means of reform and external accountability. The argument for this is that by (Tavenas 2004, p. 8):

> using objective indicators of activity, resources and performance, institutions will also be able to develop an informed and constructive dialogue with their regulatory authorities and with all partners involved in financing them. Common evaluation and Quality Assurance systems will enable them to assure the authorities of the quality of their programmes and, by the same token, of the efficiency of public investment in higher education institutions.

However, the danger inherent in evaluation policies that are based on such empiricist, objectivist, and utilitarian assumptions is that (Reichert and Tauch 2003, p. 102):

> if accountability and evaluation are reduced to a primarily technical exercise by way of rigid output measures and overly standardized evaluation exercises, then the essential debate about the values and assets which HEIs are best suited to pursue for society is clearly at risk.

That is, what has made universities great over the centuries may be compromised, if not lost.

## Advancing Quality Assurance in Engineering Education

Certainly a positive outcome of the Quality Assurance movement has been an increased emphasis on and engagement of a broad range of stakeholders in higher education. Chua (2004, p. 183) explains that higher education stakeholders understand the concept of quality in different ways.

*Parents*. Parents look at quality as relating to input (university ranking, performance, infrastructure, etc.) as well as output (employability, graduate placement, etc.).

*Students*. Students perceive quality as relating to the educational process and how they will fit in (teaching/learning, courses, etc.) as well as outputs (learning and employability).

*Faculty*. Faculty recognize quality as relating to the whole system of education and its improvement (input, process, and output).

*Employers*. Employers perceive quality in terms of the output, i.e., the ability to perform in the work place as shown through graduate attributes and competencies.

Whether it is an accountability mentality per se or a more general concern for quality, the shift in power is obvious. And, as has been the case with *EC2000,* the shift can act as a positive stimulus for improving Engineering Education. There are however, as the chapters in this book suggest, impediments to advancing Engineering Education Quality Assurance.

### *Inconsistency*

One concern regarding the advancement of Engineering Education Quality Assurance worldwide is the lack of uniformity in Accreditation standards and practices. For example, within the Washington Accord signatories, each country has individual accreditation processes and variations in accreditation criteria as well as different documentation requirements and reporting processes. In addition, in countries without a national accreditation organization the major concern for an institution is to select an appropriate accreditation body. And, there are variations in the visiting process, report writing or documentation, and assessment in these countries.

Such variations may be addressed, for example, by having visiting panels comprised of representatives from other signatory countries so that standards are maintained within the context of local variance as is the case with Washington Accord signatories. And while there are some cases where an institution can choose the agency to approach for assessment authority, in most countries it is mandatory to seek accreditation from the national accreditation agency which often has ties to a global and/or regional Quality Assurance network. In addition, ABET, Inc., the Washington Accords, EUR-ACE, and the INQAAHE provide helpful guidance for developing accreditation standards and processes. The chapters in this book by

Woollacott ("Taxonomies of Engineering Competencies and Quality Assurance in Engineering Education"), Hanrahan ("Toward Consensus Global Standards for Quality Assurance of Engineering Programmes"), and Brodeur and Crawley ("CDIO and Quality Assurance: Using the Standards for Continuous Programme Improvement") provide syntheses that may also help to foster some consistency, if not uniformity in Engineering Education Quality Assurance globally.

## *Cost*

The cost of belonging to an Accreditation agency and the fees charged for accreditation visits vary considerably. However, the greatest cost is the time and resources spent on planning and implementing a self-study and hosting a visiting team. This involves forming study teams, conducting extensive investigations, and summarizing the findings in the format specified by the Accreditation agency. And, during the visit there are transportation, room and board, and logistical costs. In many institutions these resources are simply not available or their use for Accreditation means that other essential functions are short changed.

There are also costs related to setting up internal systems and organizations, for example, institutional research and assessment management offices, to collect, analyze, and organize the information needed for Accreditation. These become fixed costs because the process of continuous improvement implied by current Accreditation standards means that assessment must become an ongoing process and not one just initiated in preparation for the next Accreditation self-study and visitation cycle.

## *Changing Expectations*

In the past, Accreditation explicitly focused on capacity, i.e., the inputs to education in the form of faculty credentials, facilities, and other infrastructure factors. Of course, processes such as curriculum and course syllabi development, budgeting and accounting practices, administrative rules and regulations, promotion and tenure procedures, admissions activities, etc. were also addressed in most accreditation standards. Explicit student learning outcomes received much less direct attention. Instead there was a general concern for the *quality* and *reputation* of graduates.

While the shift in focus to student outcomes is an additional expectation, the other input and process factors remain part of Accreditation standards. The fundamental change is that Institutional Effectiveness Assessment is intended to determine the extent to which institutional and programmatic inputs and processes foster desired learning outcomes. That is, the purpose of higher education is to provide instruction *that produces learning* and, ultimately, the test of an institution's quality is the *success of its students.*

## Conclusions

The keys to advancing Quality Assurance are to, first, strike a balance between the expectations for internal improvement and external accountability; second, recognize the value of various evaluation and assessment methods for different purposes; and, third, acknowledge the trade-offs and tensions inherent in various approaches to Quality Assurance.

The changes in institutions and programs implied by the focus on student learning outcomes and institutional effectiveness assessment must start at the most local level, i.e., individual courses or modules; majors or programs of study; colleges, departments, or divisions; and, ultimately, institutions. The task of documenting such changes and, thereby, recognizing the impact of Quality Assurance policies and practices (Accreditation and Evaluation or Assessment) makes it necessary to use different metrics at different levels of a higher education institution.

It is important to avoid the assumption that just because a definition of Quality Assurance includes improvement and accountability, that the same evaluation and assessment methods are appropriate for both purposes. These are actually two very different *ends* that require different *means* which, while not entirely separate, are quite distinct in many ways. This is where the value of the conceptual framework described in the chapter "Internal and External Quality Assurance Approaches for Improvement and Accountability: A Conceptual Framework" by Gray and Patil can be seen since it acknowledges all types of Quality Assurance approaches. The conceptual framework in Fig. 1 of the chapter "Internal and External Quality Assurance Approaches for Improvement and Accountability: A Conceptual Framework" provides a way to communicate the complexity of Quality Assurance and to adapt various approaches in a sensitive way to multiple ends and audiences.[14]

Finally, because of the different historical, philosophical, political, and social factors that have influenced Quality Assurance over the last century, there will always be tensions and conflicts because of the language used and the power implications. In many ways, Quality Assurance remains an innovation in higher education. As such, the only way to foster its adoption is to convince individual faculty members of its value through leadership, communication, involvement, and a process of planned change over a long period of time that leads to its adaptation to local conditions (Gray 1997).

---

[14] A more detailed description of such a system is beyond the scope of this chapter and will be left to the chapter "Internal and External Quality Assurance Approaches for Improvement and Accountability: A Conceptual Framework" and other venues intended to provide practical advice and direction for the development and implementation of a comprehensive Quality Assurance system.

**Acknowledgement** Peter J. Gray wishes to express appreciation to USNA Academic Dean and Provost William Miller for granting him the sabbatical leave that provided the opportunity to complete this book.

## References

AAC&U Board of Directors. (2008). *Our students best work: A Framework for accountability worthy of our mission. A statement from the board of directors* (2nd ed.). Washington, DC: Association of American Colleges and Universities. Accessed January 2009. http://www.aacu.org/publications/pdfs/StudentsBestReport.pdf

Augusti, G. (2007). Accreditation of engineering programmes at European level. *International Journal of Electrical Engineering Education, 44*(2), 101–108.

Banta, T. W., & Associates. (2002). *Building a scholarship of assessment*. San Francisco, CA: Jossey-Bass.

Brooks, N. (2005). *Taxes are the basis of civilization*. Winnipeg: Winnipeg Free Press, Friday 23 December 2005, Section: Focus, A15. Retrieved on 21 December 2008 from http://osgoode.yorku.ca/media2.nsf/5457ed39bc56dbfd852571e900728656/73eef396d217f906852570eb006f8da4!OpenDocument

CHEA Web site: A directory of CHEA recognized organizations, http://www.chea.org/Directories/index.asp and An overview of U. S. accreditation, Accessed January 2009. http://www.chea.org/pdf/overview_US_accred_8-03.pdf

Cheng, Y., & Liu, N. C. (2008). Examining major rankings according to the Berlin Principles. *Higher Education in Europe, 33*(2/3), 201–208. Online Publication Date: 01 July 2008.

Chua, C. (2004). Perception of quality in higher education. In R. Carmichael (Ed.), *Proceedings of the Australian Universities Quality Forum 2004, Adelaide, Australia, 7–9 July 2004* (pp. 181–186). Melbourne: Australian Universities Quality Agency (AUQA).

Clarke, M. (2007). The impact of higher education rankings on student access, Choice, and Opportunity. In Institute for Higher Education Policy (Ed.) *College and university ranking systems: Global perspectives and American challenges* (pp. 35–48). Washington, DC: Institute for Higher Education Policy.

CTI (La Commission des Titres de I'Ingénieur). (2006). Accessed 18 September 2006. http://www.cefi.org/CEFINET/GLOBAL/CTI/TITRE_2/INDEX.HTM

East, D. (2008). *Understanding institutional performance: Advice to the secretary of state for innovation, universities and skills*. Bristol: Higher Education Funding Council for England (HEFCE).

Eaton, J. S. (2008). Higher Education Opportunity Act of 2008: What does it mean and what does it do? (first of two articles). Inside Accreditation with the President of CHEA 4, 1. http://www.chea.org/ia/IA_2008.10.30.html. Accessed January 2009.

ENQA. (2007). *European network for quality assurance in higher education*. Accessed 2007. http://www.enqa.eu/

ESOEPE. (2005). European Standing Observatory for the Engineering Profession and Education (ESOEPE), Definitions/Glossary, FEANI definition (after discussion at the PSC Meeting of 11 May 2001). Accessed January 2009. http://www.feani.org/ESOEPE/HomePage.htm

Ewell, P. T. (1989). Hearts and minds: Some reflections on the ideologies of assessment. In: *Three Presentations from the Fourth National Conference on Assessment in Higher Education* (pp. 1–26). Washington, DC: American Association of Higher Education.

Ewell, P. T. (1991). To capture the ineffable: New forms of assessment in higher education. *Review of Research in Education, 17*, 75–125.

Gray, P. J. (1997). Viewing assessment as an innovation: Leadership and the change process. In T. W. Banta & P. J. Gray (Eds.), *The campus-level impact of assessment: Progress, problems, and possibilities. New directions in higher education* (Vol. 100, pp. 5–15) San Fransisco, CA: Jossey-Bass.

Gray, P. J. (2002). The roots of assessment: Tensions, solutions, and research directions. In T. W. Banta (Ed.), *The scholarship of assessment: What are we learning from assessment?* (pp. 49–66). San Francisco, CA: Jossey-Bass.

Harman, G. S., & Meek, V. L. (2000). *Repositioning quality assurance and accreditation in Australian higher education*. Canberra: Canberra Commonwealth of Australia, Department of Education Training and Youth Affairs (DETYA). Evaluations and Investigations Programme Higher Education Division, AusInfo.

Hoecht, A. (2006). Quality assurance in UK higher education: Issues of trust, control, professional autonomy and accountability. *Higher Education, 51*(4), 541–563.

International Engineering Alliance. (2007). *International educational accords: Rules and procedures resulting from International Engineering Meeting (IEM) Washington, Wednesday 15th Aug 2007*. http://www.washingtonaccord.org/Rules-and-Procedures-Aug-2007.pdf. Washington Accord 1989: http://www.washingtonaccord.org/Washington-Accord/; Sydney Accord 2001: http://www.washingtonaccord.org/sydney/; Dublin Accord 2002: http://www.washingtonaccord.org/Dublin/; Accessed January 2009.

Irandoust, S., Nicklasson, C., & Sjöberg, J. (2000). Do national quality audits enhance quality at the institutional level? Experience from Chalmers University of Technology. *Proceedings of the 2nd Global Congress on Engineering Education, Wismar, Germany* (pp. 261–273).

Joint Quality Initiative. (2004). Shared 'Dublin' descriptors for short cycle, first cycle, second cycle and third cycle awards. Draft 1 working document on JQI meeting in Dublin on 18 October 2004. Accessed January 2009. http://www.eua.be/fileadmin/user_upload/files/EUA1_documents/dublin_descriptors.pdf

Lederman, D. (2009). *A call for assessment – Of the right kind*. Inside Higher Ed.com, January 8 2009. Accessed January 2009. http://www.insidehighered.com/news/2009/01/08/aacu

López-Segrera, F. (2007). The approach to the concepts of quality and accreditation in the UNESCO World Conference on Higher Education (Paris, 1998) and follow-up meetings. In *Accreditation for Quality Assurance: What is at stake?* (2nd ed., pp. xlvi–xlviii). GUNI Series on the Social Commitment of Universities, Global University Network for Innovation. Hampshire: Palgrave Macmillan.

Marchese, T. J. (1987). Assessment update: Third down, ten years to go. *AAHE Bulletin, 40*(4), 3–8. Eric Document: #301120.

Merwin, J. C. (1969). Historical review of changing conceptions of evaluation. In R. W. Tyler (Ed.), *68th NSSE yearbook: Educational evaluation: New roles, new means* (pp. 6–25). Chicago, IL: National Society for the Study of Education (NSSE), University of Chicago Press.

Neal-Sturgess, C. (2007). Bologna and the MEng: 'Sleepwalking into unknown and unpredictable territory'. *International Journal of Electrical Engineering Education, 44*(2), 129–130. Accessed January 2009. http://journals.mup.man.ac.uk/cgi-bin/pdfdisp//MUPpdf/IJEEE/V44I2/440129.pdf

New York State Education Department Office of Higher Education Web site: http://www.highered.nysed.gov/ocue/; Board of Regents Authority http://www.highered.nysed.gov/ocue/board_of_regents_authority_for_q.htm Accessed January 2009.

North Carolina State University. (2009). *Internet resources for higher education outcomes assessment*. Raleigh, NC: NC State University. Planning & Analysis Web site: http://www2.acs.ncsu.edu/UPA/assmt/resource.htm#toc. Accessed January 2009.

Palomba, C. A., & Banta, T. W. (1999). *Assessment essentials: Planning, implementing, and improving assessment in higher education*. San Francisco, CA: Jossey-Bass.

Patil, A., & Codner, G. (2007). Accreditation of engineering education: review, observations and proposal for global accreditation. *European Journal of Engineering Education, 32*, 639–651.

Patil, A., & Codner, G. (2008). Accreditation in engineering education: Findings from selected Asia-Pacific countries. In J. Steinbach (Ed.), *Proceedings of the 2nd Deans Conference – Special Challenges in Engineering Education, 2008*. European Society for Engineering Education (SEFI): Berlin, Germany.

Popham, J., & Baker, E. (1970). *Systematic instruction*. Englewood Cliffs, NJ: Prentice Hall.

Reichert. S. (2008). Looking back – looking forward: Quality assurance and the Bologna process. In: A. Beso, L. Bollaert, B. Curvale, H. T. Jensen, L. Harvey, E. Helle, B. Maguire, A. Mikkola and A. Sursock (Eds.) *Implementing and using quality assurance strategy and practice*. A selection of papers from the 2nd European Quality Assurance Forum Hosted by Sapienza Universita Roma, 15–17 November 2007. Brussels, Belgium: The European University Association.

Reichert, S., & Tauch, C. (2003). *Trends in learning structures in European higher education III – Bologna four years after: Steps towards sustainable reform of higher education in Europe, First draft*. Graz: European University Association; European Commission.

Sadlak, J., Merisotis, J., & Liu, N. C. (2008). University rankings: Seeking prestige, raising visibility and embedding quality – the Editors views. *Higher Education in Europe, 33*(2), 195–199.

Schwarz, S., & Westerheijden, D. F. (2007). Accreditation in the framework of evaluation activities: A comparative study in the European Higher Education Area. In S. Schwarz & D. F. Westerheijden (Eds.), *Accreditation and evaluation in the European higher education area*. Dordrecht: Springer.

Scriven, M. (1967). The methodology of evaluation. In R. W. Tyler, R. M. Gagné & M. Scriven (Eds.), *Perspectives of curriculum evaluation* (pp. 39–83). Chicago, IL: Rand McNally.

Stake, R. E. (1967). The countenance of educational evaluation. *Teachers College Record, 68*, 523–540.

Stake. R. E. (1975). *Program evaluation particularly responsive evaluation*. Occasional Paper Series, Paper #5. Urbana-Champaign, IL: University of Illinois at Urbana-Champaign, Center for Instructional Research and Curriculum Evaluation.

Stake, R. E. (1991). Retrospective on the countenance of educational evaluation. In M. W. McLaughlin & D. C. Phillips (Eds.), *Evaluation and education: At quarter century. Ninth yearbook of the national society for the study of education*. Chicago, IL: University of Chicago Press.

Stufflebeam, D. L., Foley, N. J., Gephart, W. J., Guba, E. G., Merriman, H. O., & Provus, M. M. (1971). *Educational evaluation and decision making*. Itasca, IL: Peacock Publishers.

Tavenas. F. (2004). *Quality assurance: A reference system for indicators and evaluation procedures*. Brussels: European University Association. Accessed January 2009. A free electronic version of this report is available through http://www.eua.be

The Quality Assurance Agency for Higher Education. (2006). *Code of practice for the assurance of academic quality and standards in higher education, Section 6: Assessment of students – September 2006* (2nd ed.). Accessed January 2009. http://www.qaa.ac.uk/academicinfrastructure/codeOfPractice/section6/default.asp

Usher, A., & Savino. M. A. (2007). Global survey of rankings and league tables. In: Institute for Higher Education Policy (Ed.), *college and university ranking systems: Global perspectives and American challenges* (pp. 23–34). Washington, DC: Institute for Higher Education Policy.

Voluntary System of Accountability. (2009). Accessed January 2009. http://www.voluntarysystem.org/index.cfm

Woodhouse, D. (2007). *INQAAHE guidelines of good practice in quality assurance. International network for quality assurance agencies in higher education*. Accessed January 2009. http://www.inqaahe.org/Upload/Upload/INQAAHE_documents/INQAAHE_-_Guidelines_of_Good_Practice.pdf

Worthen, B. R., & Sanders, J. R. (1973). *Educational evaluation theory and practice*. Worthington, OH: Charles A Jones.

# Issues and Challenges (Global/Regional Perspectives)

# Quality Assurance in European Engineering Education: Present and Future Challenges

**John Cowan**

**Abstract** This chapter focuses on the Quality Assurance (QA) of higher engineering education in UK and Europe, by considering eight challenges which are predicted by the writer to be of increasing importance in the years ahead. QA in higher education is taken here as a process that sets out to assure society, and responsible bodies within it, about the quality of educational provision for students. The purpose of the chapter is to identify the present and forthcoming challenges and changes in QA in engineering education in UK and Europe, in the light of present circumstances as well as of the historical context.

## Introduction

There are many challenges which will become increasingly important in the years ahead for those in engineering education in Europe. This chapter is organised around the following topics, for each of which is advanced a constructive suggestion for action or a prediction of forthcoming change. All of these topics involve significant challenges as follows:

1. Responding in our Quality Assurance (QA) to political decisions seeking a unified European approach to higher education.
2. Developing the rigour of the practice of QA in engineering education.
3. Confronting the long-established practice of concentrating on relatively lower-level outcomes and aims in much of engineering education.
4. Finding effective ways to develop higher-level abilities, both cognitive and interpersonal, and to evaluate how well that is being done.
5. Arranging QA to cope with the sometimes conflicting demands of professional bodies and educational authorities.

J. Cowan (✉)
Academic Development, Edinburgh Napier University, Bevan Villa, Craighouse Campus, Edinburgh, EH10 5LG, Scotland
e-mail: J.Cowan@napier.ac.uk

A.S. Patil and P.J. Gray (eds.), *Engineering Education Quality Assurance: A Global Perspective*,
DOI 10.1007/978-1-4419-0555-0_2, 

6. Ensuring that the oversight of quality is informed, independent and objective.
7. Enabling change in some of the outmoded, but enduring, teaching practices of yesteryear.
8. Extending our QA to cover self-managed and self-directed continuing professional development (CPD).

While readers in North America and elsewhere will no doubt see striking contrasts between values, practices and trends on both sides of the Atlantic (Heywood 2005), many of the above topics relate equally to education and accreditation in other nations and professional areas. This certainly applies within Europe, as the amplification of Challenge 1 should make clear.

In considering the QA of higher engineering education, it is important to distinguish between academic awards that testify to a certain level and scope of learning and development on the part of an individual, and what is called their professional accreditation, which entitles the accredited person to practise professionally. The author will follow the predominant (but not consistent) UK usage and take assessment (Heywood 2000) to be a process in which judgements of a student's ability or understanding are made, in contrast to evaluation (Calder 1994), which is a process in which judgements are made of the standard and quality of an academic programme, or a component of it. QA is therefore an evaluative process in which consideration is given, inter alia, to the validity, reliability, relevance and standard of embedded processes of assessment.

## Challenge 1: Bologna and Thereafter

In 1999 the European Community agreed, and declared in the Bologna Declaration (European Higher Education Area 1999), that in order to promote the European system of higher education world-wide, European countries would:

- Adopt a system of easily readable and comparable degrees to promote European citizens' employability and the international competitiveness of the European higher education system.
- Adopt a system essentially based on two main cycles: undergraduate and graduate.
- Establish a system of credits (European Commission 2005) – such as in the European Credit Transfer system (ECTS) – to promote widespread student mobility.
- Promote that mobility by overcoming obstacles to the effective exercise of free movement of students, teachers, researchers and administrative staff.
- Promote European co-operation in QA, with a view to developing comparable criteria and methodologies. (Joint Quality Initiative, 2004)
- Promote the necessary European dimensions in higher education, particularly with regard to curricular development, inter-institutional co-operation, mobility schemes and integrated programmes of study, training and research.

Note that in all of the above items, general conformity across the board, and in all discipline areas, is clearly assumed. There are unlikely to be any exclusion for either a discipline or a country. Notice also though that the difference between the rhetoric of much QA procedures as documented, and reality as QA is presently practised, can be stark.

It is, of course, the penultimate bullet point which is of particular importance in the present context. The key word, perhaps, is comparable – which does not necessarily mean identical. I foresee tension between those countries in which, at present, detailed syllabi and precise standards are determined by ministries of education, and those in which the sector is virtually self-regulating, while being accountable, somehow, for its management of quality. I would expect those in the latter group to co-operate nationally and internationally as suggested, at least in the interim stages, and to do so overtly or covertly. They will strive for various reasons to bring their present systems for QA reasonably into line. However, the former group may well resist the imposition of many such standards and methodologies, because they will call for overmuch change, or will not receive ministerial approval. In addition, the professional bodies in some countries will also contribute part of the resistance to change, through fear of losing their power to determine the nature of the degrees which they accredit.

## Challenge 2: Ensuring Adequate Rigour in Quality Assurance

As an engineering academic for the last 45 years, I have had extensive and recent experience as an external examiner in the UK system (Lewis 2005) on various engineering degrees, as an international educational consultant and as an auditor/reviewer for the UK Quality Assurance Agency (Hodgson 2005). Sadly, this experience suggests that the rigour of QA in European engineering education has compared unfavourably in the past four decades with that which could be found contemporaneously in at least some other discipline areas. There is much work to be done to bring engineering in line with best practice. We need to catch up; and thereafter to progress, as some others are even now doing.

QA procedures in the more progressive educational institutions now routinely depend upon the following features of the programmes whose quality is being assured:

- Comprehensive specifications for modules or other elements of programmes (QAA 2008).
- Full alignment (Biggs 2003; Cowan 2004a) between intended learning outcomes, methods of assessment and the learning and teaching activities.
- Regular self-evaluations of programmes.
- Annual reviews of these self-evaluations by internal colleagues, drawn from outwith the programme team.
- Formal validations before first delivery of a programme, and subsequently at perhaps 5-year intervals, by panels which include external experts.
- Use of data covering both student learning and their learning experience.

- Students' involvement in the undertaking of reviews and reports.

Increasingly, in the more advanced institutions, the end result of their internal processes of review is an objective self-evaluation, formulated against declared criteria and using recognised sources of data. When these are available, it only remains for an external QA process to audit, which properly then entails confirming the adequacy and accuracy of all the elements of the internal evaluation which is placed before it – including the final internal judgement.

Such schemes for QA represent best current practice. They have only emerged and developed gradually in recent years. They are by no means the norm at the time of writing, even in the more advanced European countries. But they occur sufficiently frequently to demonstrate that the above features are feasible, and are of benefit (eventually, perhaps) to the institution and to the country. Consequently they will increasingly encourage those in authority elsewhere to expect, if not demand, such activity of their staff.

## Challenge 3: Level of Expected Outcomes

Forty years ago, much of engineering education, if judged by its examinations and coursework, concentrated upon the assimilation, understanding and recall of basic knowledge, together with the application of routine algorithms or methods to carry out somewhat predictable calculations (Cowan 2006a). The higher-level abilities of analysis, creativity and synthesis, and the making of judgements, seemed to be expected to develop by osmosis (Bowden 2004). Interpersonal skills, which figure so highly in our professional lives, were often not touched developmentally in formal curricula (Cowan 2004b).

Nowadays, the situation has changed radically. In professional practice, the routine knowledge which engineers require can usually be retrieved in a suitable form through the simple use of a search engine. Explanations, if needed, can be similarly accessed, and need only be mastered when they are needed. Routine applications are readily undertaken on our behalves by commercial software (Cowan 2006b). Yet it is my experience, particularly as an external examiner and auditor, that many of the demands in current day engineering assessments are still at a regrettably low and inappropriate level. They are thus often redundant in terms of usefulness in employment after graduation.[1]

---

[1] A former student of mine, who has had extensive experience in engineering education and is now a well-regarded member of senior academic management, kindly read an earlier draft of this paper and commented (Matthew 2008) in support of this point that:

> From my experience of engineering education, your challenges 3, 4 & 5 are the key ones – and there seems to be a real conflict here between what the professional bodies say in accreditation literature and the reality of what they look for on accreditation visits to university departments. My experience leads me to think there is still an undue emphasis on the low level, easy to measure abilities and skills and little pressure put on university departments to radically change the curricula and the pedagogy to really deliver the kind of engineers that the professional bodies profess to want.

QA and external examining procedures in universities have become increasingly aware of this weakness, and are calling for attention to be given to it. In particular we urgently need, as a profession, to align our assessed demands, the so-called hidden curriculum (Snyder 1971), with the requirements of the profession and the expectations of employers. That need is easier stated than achieved. However, failure to deal with it may prove a stick for our backs. For the management of QA increasingly adopts a cross-disciplinary approach and many powerful personalities nowadays wish to ensure comparable levels of demand across disciplines and their awards. This trend, which naturally leads to comparisons and consequent criticisms, is particularly apparent in European discussions and negotiations subsequent to the Bologna Agreement, as nations have sought to face up to its implications (see Challenge 1).

## Challenge 4: Developing, Assessing and Evaluating Higher-Level Abilities

When I began to teach structural engineering in 1964, it was easy to confirm that a student understood a concept; we could simply ask them to explain it. It was relatively straightforward to teach towards that understanding. We could explain, and give examples, until the concept had been grasped. It was equally straightforward, having assessed the student's grasp of that concept, to then assess their ability to apply that understanding, in particular examples.

It is considerably more difficult to bring about achievement of today's higher-level educational demands. In our teaching nowadays we should be developing in students their ability to generate creative solutions in problem solving (Cowan 2006c). We should also be developing their ability to expand their original plans in detail, and then to judge the merit of these solutions, comparatively and objectively. These are demanding pedagogical challenges. They call on us to create and deliver effective learning and teaching activities and to have confidence in their outcomes. It is yet more demanding to work out how to assess the development of these abilities in our examinations and assignments. And it is even more difficult, for those who are responsible for QA, to make and confirm judgements on the effectiveness of such learning and teaching activities and on the alignment of the assessment instruments. In European practice generally, this is an important challenge with which little progress has been made at the time of writing, and even that merely in pockets of specialised activity.

It seems likely that developments in QA will depend upon the involvement of students as colleagues in programme evaluation and QA (Brooker and Macdonald 1999). They can assist, for a start, in determining what development of abilities has actually taken place (Campbell et al. 2007). This would be especially true of the involvement of recent former students, now in the market place, and who have gained a reflective perspective into the curriculum, For example, it is often only the learners themselves who know, and can claim objectively, the true extent of their creativity (Cowan 2006c). Already in some jurisdictions, including my own country of Scotland, we are seeing the active involvement and integration of trained students in QA procedures. These scrutinise the effectiveness and standard of provision (Gordon 2002), whose outcomes feature inter alia higher-level learning and development – and focus in addition on enhancement-led review. There appears considerable potential in this latter development in matters of quality, although already there is perhaps an increasing danger (Matthew 2008) in that some people want to quality-assure quality enhancements, rather than systematically evaluate their impact on the student experience.

## Challenge 5: Integrating Conflicting Demands

Traditionally, professional bodies (Maillardet 2004) have (rightly and understandably) concentrated upon ensuring that graduates have the necessary grasp of disciplinary fundamentals (Heywood 2005), together with proven competence in essential professional skills – before they seek licence to practice. Graduates should then progress to accreditation and professional status, by demonstrating that they have had suitable practical experience and have developed necessary practical competences (Becher 1999).

Until recently, educational institutions have found little difficulty in responding to these expectations of their role in the first stage of this process. They have internally validated their programmes, and confidently exposed them to a second stage in which these programmes are accepted by the relevant professional body or bodies, or even, in some countries, by government ministries.

Recently, however, problems have troubled this arrangement. These have arisen because:

- Developments based on information technology have removed many lower-level skills from curricula (see Challenge 4). Many of the basic engineering skills are now economically and more effectively delivered by the new technology. In their place employers therefore rightly look for employable graduates to offer higher-level cognitive, interpersonal and social skills (Beder 1999). These are generic rather than disciplinary, and should apply and be developed across our curricula.
- As already mentioned, the subject matter of engineering courses has an increasingly short shelf-life. The content which students study in an up-to-date programme will be partly out of date before they can apply it in practice. Mastery of subject matter which has only emerged since they graduated will be required of them (see Challenge 8).
- During their professional practice, graduates will then have to engage responsibly and effectively with their own professional development, both immediate and long-term; higher education must therefore devote time (taken from disciplinary subject coverage) to equipping them with the skills for self-directed lifelong learning (Candy 1991).

Consequently assurance of the quality, an overall process leading to a licence to practice, must now cover mastery of content encountered after graduation coupled with a reasonable assurance of the ability to master new content and skills. To this demand should be added the complication that the processes of professional bodies have, in the main, been self-assured (see Challenge 5, below).

At least one professional body outwith engineering, namely the Chartered Institute of Personnel and Development, has dealt with this in an imaginative way (Chartered Institution of Personnel and Development 2005). This body recognises that its QA procedures, as a professional body, cannot cover all that is required if there is to be thorough oversight of its professional accreditation. They therefore collaborate with universities in the provision of supplementary activities, external to degree programmes – for example, in human resources management. The Institute specifies the coverage, standard and assessment of such provision – and remits it, in partnership, to a collaborating university's QA procedures to cover the additional professional provision, as well as the academic degree programme (Francis and Cowan 2008).

Is this not a possible way ahead for engineering education? The learning and development required for professional accreditation go beyond the coverage of an undergraduate degree. The assessment of this should surely be left, as before, with the professional body. However, the QA of the total process should be a holistic confirmation of quality, probably integrated with the procedures of the university, while remaining open to scrutiny by the professional body. In other words, I advocate and forecast within Europe an integration of QA procedures for both degree programmes and professional validation.

## Challenge 6: Ensuring Informed, Independent and Objective Oversight

Before we had any procedures for QA, it was common to judge personal or group teaching performance against somewhat vague criteria, which were personally determined or set by individuals or a programme team. Usually this activity was based merely on impressions of the situation being judged, rather than on objective data assembled to describe the situation and learning outcomes.

The subsequent development of QA approaches (Harvey 2005) has arisen from reasonable reservations about this process, which thoughtful observers and participants had formulated. They noted commonly occurring situations in which those who planned and delivered and assessed programmes also acted, in effect, as custodians of their own standards. The move towards objective self-evaluation has certainly been accompanied by an expectation that criteria and sources of data will be declared, explicit, and transparent. It has also been followed by the (reasonable) view that externality, at least internal externality, is desirable, when judgements are being formulated.[2]

We still have some way to go before externality is specified as an essential feature of quality reporting and of review. Beyond that goal, we will then need to ensure that those who contribute to review as externals are adequately trained or prepared to follow, and if necessary insist upon, an objective process. This is perhaps especially so in the case of an international dimension to externality, both with the envisaged establishment of international agencies, in the context of Bologna, and also with the fact that in small countries or specialist disciplines, in which everybody in HE knows almost everybody else, competent and independent externality will be an important and desirable feature of assurance.

## Challenge 7: Increasing Educational Professionalism

A generation ago, those who taught engineering were usually professionally qualified – in their discipline of engineering. But the notion of being professionally qualified to plan provision, to deliver teaching and to assess, was seldom aired. Some academics concentrated upon research or consultancy; the majority, in their teaching, relied on their own past experience and common sense, often merely justified as meaningful *gut reactions*, which they would have been hard pressed to distinguish from indigestion.

There followed perhaps 15 years of gentle transition, in reaction in mainland Europe to the students' revolts of 1968. Project-oriented learning often replaced didactic instruction (Kjersdam and Enemark 1994). A minority of teachers were minded to develop their teaching founded upon an acquaintance with basic research findings regarding pedagogy. A few enthusiasts and visionaries began to offer what they called *freedom in learning* (Rogers 1969) or *independent learning* (Robbins 1988). And in some universities in some countries, attempts were made to provide training for newly appointed or otherwise interested lecturers in the areas of teaching, learning and assessment.

---

[2]I take externality in quality judgements to mean the primary involvement therein of persons who are external to the programme or activity whose quality and standards are being judged. I take internal externality to describe the usefully constructive process by which judgements are made by colleagues in the same institution, but drawn from different discipline areas. External externality involves at least some completely independent panel members, who bring even greater detachment and useful breadth of experiences to the process.

Inexorably external agencies were required by society and established initially to judge the quality and standard of programmes in higher education. Nowadays they are more likely to have to scrutinise the manner in which the institution satisfies itself with regard to the standard of its awards, and the quality of the learning experience it provides. Programme teams and disciplinary schools have naturally become increasingly adept at tactically assembling data, or fragments of data, which can influence visiting panels to form favourable judgements. In response, zealous auditing teams, internal as well as external, have acquired skills of probing enquiry, to ensure balanced and rigourous judgements!

In Britain, by 1990, most universities were expecting new appointees to undertake training. A consequence of the Dearing Report (HMSO 1997) to government, and the decisions arising from that, has been that induction training will become mandatory in UK, around the time of writing. This has had a noteworthy effect on the pedagogical knowledge base which now informs curriculum development, review and QA. Lecturers are expected nowadays to have engaged and continue to engage proactively with the Scholarship of Learning and Teaching (Boyer 1990, 1998), as it has been called.

Each year, a further cohort of moderately revolutionary Young Turks emerges from training programmes into practice. They have qualifications and engage in creative thinking about their curricula, which has been stimulated by their studies for postgraduate certificates in higher education. Increasingly they are a strong, dominant and informed voice in decision-making groups. Additionally, each year, the inexorable march of time brings about the professional demise of some of the Old Guard. For a changing of the guard inevitably occurs with their retirement, removing much diehard educational conservatism in consequence.

This progression is tangible, and is now by no means slow. Since 1990, publications on staff and curriculum development have rapidly become more and more professional, more based upon properly evaluated pilots and formulated theories, and less on anecdotal accounts of innovation accompanied by enthusiastic endorsement from the innovators. Higher education is therefore fast earning itself the right to be regarded as a professional practice. It is increasingly based, just as a profession such as engineering should be, on familiarity with a sound knowledge base, on generally accepted and proven practice and on developments emerging from ongoing research (Rushby and Cowan 2006).

However, a new hazard to quality and standards has emerged. This challenge presents an interesting dilemma – in that whilst many who teach now have training in teaching, increasingly engineering departments are staffed by non-engineers, or at least by some without professional experience or qualifications. So what is the impact of this on the quality of engineering education presented? This becomes a really important issue in the area of design education, where many of the staff may have no engineering design experience.

Inevitably, QA activities in the future will also be increasingly founded upon the professional base of our new discipline of higher education, yet engineering education should surely still depend on the professional competence of staff as engineers. QA will prompt development of both aspects of that base and the enhancement (Raban 2007) of HE provision. It will do so with agenda items arising from questions, issues and examples of good practice which are identified during QA activities, and international scrutiny of these, arising in consequence.[3]

[3] A European colleague commented (Oliveira 2008) that in the last few paragraphs of this section, I concentrate on the British QA reality. It is his belief that a brief view of what is happening, or not yet happening, in the rest of Europe would illustrate how much diversity exists, and that Britain is probably years ahead of much of the rest of Europe. I concur, but would not wish to make invidious comparisons here in any detail.

## Challenge 8: Continuing Professional Development

One consequence of the information explosion has been recognition in most professional areas of the need to ensure that practising professionals continue to undertake adequate professional development. They should update their knowledge and skills, and even uprate them. However, most of the arrangements which have been made to date with this end in mind are somewhat suspect in terms of assured quality. It is common nowadays for professionals to maintain a record of their attendance at CPD events or of other activity which they claim has contributed to their development. Yet even certificates of attendance (commonly issued and retained) do not certify that the attendees were awake or attentive during the session. They certainly do not attest to retained learning or development, which is what important in worthwhile CPD.

It is rare – very rare at present – for any check to be made of the standard of learning and development claimed in CPD, or of the effectiveness of the learning or developmental experience. Yet these features are now basic and vital constituents of a QA approach to formal graduate education. It seems likely, and highly justifiable, that a society which looks for QA of the education provided by universities, should soon expect a similar oversight in respect of CPD, whatever provider or manager is involved. In similar vein, society also remits to us the recognition and accreditation of Prior Learning, a much talked about issue, whose practices lack insight, experience, consistency and rigour, and which therefore should also be subject to QA procedures. I would hope that this will be yet another example of a feature in which post-Bologna comparisons will lead to the identification of discrepancies judged to be important, with the consequence that pressure will be brought to bear on weaker providers (and national practices).

## Conclusions or Predictions

From the thoughts I have set out here, I am suggesting that the future, as far as QA in engineering education in Europe is concerned, will bring:

- Greater and more consistent rigour in QA processes
- More emphasis on providing effective teaching for the attainment of higher-level learning outcomes
- The development of sound methods of assuring the quality and standards of the attainment of higher-level learning outcomes, both cognitive and interpersonal and including professional competence
- Externality becoming an accepted and routine feature of the reviews and audits in QA
- QA activities which will build upon the professional base of our new discipline of higher education, and which will prompt development and enhancement.
- Forceful efforts by the European Community to establish comparable criteria and methodologies for QA in higher education, which will have a powerful

impact on generic requirements that were not designed with engineering specifically in mind.
- Steps to tackle the demanding challenge of assuring the quality of the CPD which so many professions now require – and accredit.

I envisage these as changes in the future, though certainly not in all cases in the near future! Nevertheless, as my Portuguese colleague (Oliveira 2008) wisely points out, QA in Europe may well be regarded not only as a tool for transparency and mobility, but also as a tool for the reform of European higher education, as envisioned in the Bologna Declaration.

## References


Becher, T. (1999). Quality in the professions. *Studies in Higher Education, 24*(2), 225–235.

Beder, S. (1999). *Beyond technicalities: Expanding engineering thinking*. Faculty of Arts – Papers, University of Wollongong.

Biggs, J. (2003). *Teaching for quality learning at university*. Maidenhead: Open University Press.

Bowden, J. A. (2004). Capabilities-driven curriculum design. In C. Baillie & I. Moore (Eds.), *Effective learning and teaching in engineering*. Oxon: RoutledgeFalmer.

Boyer, E. L. (1990). *Scholarship revisited*. Princeton, NJ: Princeton University, Carnegie Foundation for the Advancement of Teaching.

Boyer, E. L. (1998). *Reinventing undergraduate education: A blueprint for America's research universities*. Stony Brook, NY: State University of New York.

Brooker, R., & Macdonald, D. (1999). Did we hear you? Issues of student voice in a curriculum innovation. *Journal of Curriculum Studies, 31*(1), 83–97.

Calder, J. (1994). *Programme evaluation and quality*. London: Kogan Page

Campbell, F., Beasley, L., Eland, J., & Rumpus, A. (2007). *Hearing the student voice*. Edinburgh: Napier University, HEA Education Subject Centre.

Candy, P. C. (1991). *Self-direction for lifelong learning*. San Francisco, CA: Jossey Bass.

Chartered Institution of Personnel and Development. (2005). *Fresh thinking on CPD*. Available at: http://www.cipd.co.uk/mandq/routes/educate

Cowan, J. (2004a). Education for higher level abilities; Beyond alignment to integration. In V. M. S. Gil, I. Alarcão & H. Hooghoff (Eds.), *Challenges in teaching and learning in higher education*. Aveiro: University of Aveiro.

Cowan, J. (2004b). Beyond reflection: Where next for curricula which concentrate on abilities? In C. Baillie & I. Moore (Eds.), *Effective learning and teaching in engineering*. Oxon: Routledge Falmer.

Cowan, J. (2006a). *On becoming an innovative university teacher*. Maidenhead: Open University Press.

Cowan, J. (2006b). In J. O'Donoghue (Ed.), *Introduction to technology supported learning and teaching*. Hershey: Information Science Publishing.

Cowan, J. (2006c). How should I assess creativity? In N. Jackson, M. Oliver, M. Shaw & J. Wisdom (Eds.), *Developing creativity in higher education*. London: Routledge.

European Commission. (2005). *ECTS users' guide*. Brussels: Directorate-General for Education and Culture.

European Higher Education Area. (1999). *Joint declaration of the European Ministers of Education*. Available at: http://www.bologna-bergen2005.no/Docs

Francis, H., & Cowan, J. (2008). Fostering an action-reflection dynamic amongst student practitioners. *Journal of European Industrial Training, 32*(5), 336–346.

Gordon, G. (2002). Learning from quality assessment. In S. Ketteridge, S. Marshall & H. Fry (Eds.), *The effective academic*. London: Kogan Page.

Harvey, L. (2005). A history and critique of quality evaluation in the UK. *Quality Assurance in Education, 13*(4), 263–276.
Heywood, J. (2000). *Assessment in higher education*. London: Jessica Kingsley.
Heywood, J. (2005). *Engineering education*. Hoboken: IEEE Press.
HMSO. (1997). *National Committee of Inquiry into Higher Education (the Dearing Report)*. London: Her Majesty's Stationery Office (HMSO).
Hodgson, J. (2005). The proposed new method of QAA review: The saga continues. *ALT Bulletin* 109. London: The Association of Law Teachers. http://www.qaa.ac.uk/academicinfrastructure/codeOfPractice/default.asp
Joint Quality Initiative. (2004). *Shared "Dublin" descriptors for short cycle, first cycle, second cycle and third cycle awards*. http://www.jointquality.nl/
Kjersdam, F., & Enemark, S. (1994). *The Aalborg experiment – Project innovation in university education*. Denmark: Aalborg University Press.
Lewis, R. (2005). External Examiner System in the United Kingdom: Fresh Challenges to an Old System on http://www.unc.edu/ppaq/docs/ExExaminers.pdf
Maillardet, F. (2004). What outcomes is engineering education trying to achieve? In C. Baillie & I. Moore (Eds.), *Effective learning and teaching in engineering*. Oxon: RoutledgeFalmer.
Matthew, R. (2008). Personal communication, when asked to comment upon an early draft of this chapter, as an engineer and Vice-Principal (Learning and Teaching) of the UHI Millennium Institute.
Oliveira, J. M. N. (2008). Private communication, having considered the chapter from a Portuguese standpoint, as team leader for the introduction of POPBL degrees in engineering.
QAA (Quality Assurance Agency for Higher Education). (2008). *Code of practice for the assurance of academic quality and standards in higher education*. http://www.qaa.ac.uk/academicinfrastructure/codeofpractice/
Raban, C. (2007). Assurance versus enhancement: Less is more? *Journal of Further and Higher Education, 31*(1), 77–85.
Robbins, D. (1988). *The rise of independent study: The politics and the philosophy of an educational innovation, 1970–87*. Maidenhead: Open University Press.
Rogers, C. R. (1969). *Freedom to learn*. Boston, MA: Houghton Mifflin.
Rushby, N., & Cowan, J. (2006). The quality of research. *British Journal of Educational Technology, 37*(5), 659–663.
Snyder, B. R. (1971). *The hidden curriculum*. New York, NY: Knopf.

# EUR-ACE: The European Accreditation System of Engineering Education and Its Global Context

**Giuliano Augusti**

**Abstract** The *EUR*opean *AC*redited *E*ngineer (EUR-ACE) project (2004/06) formulated Framework Standards for the European Accreditation of Higher Education Programs in Engineering. The EUR-ACE accreditation system is now being implemented. The European Network for Accreditation of Engineering Education has been established to run the system and six agencies have been accredited and have started awarding the EUR-ACE label in six countries (France, Germany, Ireland, Portugal, Russia, and UK). Contacts are also in progress with accrediting agencies outside the European Higher Education Area.

## Introduction

Accreditation of engineering educational programs as an entry route to the engineering profession has proved to be a powerful tool to improve both academic quality and relevance for the job market (Augusti et al. 2007). Indeed, the word accreditation, used in the United States since the 1930s, did not find its way into European specialized literature and official documents until recently: however, historically Europe has been in the forefront of such efforts.

Within continental Europe, formal accreditation (habilitation) was started in France. A 1934 law established the *Commission des Titres d' Ingénieur* (CTI), in which not only academia but also employers and social stakeholders are represented on a parity basis. Only graduates from a program with the CTI habilitation can use the title of *ingénieur diplômé*. At present, some 700 engineering programs are accredited in French schools. In UK a similar role has been played since the nineteenth century by the Professional Institutions of the different engineering disciplines (branches). These institutions exempted the graduates of accredited higher education programs from some professional admission requirements. As a result, in UK accreditation is distinguished by discipline. In 1981 the Engineering

G. Augusti (✉)
Università La Sapienza – Facoltà di Ingegneria, Via Eudossiana 18, I-00184 Roma, Italy
e-mail: giuliano.augusti@uniroma1.it

A.S. Patil and P.J. Gray (eds.), *Engineering Education Quality Assurance: A Global Perspective*,
DOI 10.1007/978-1-4419-0555-0_3, © Springer Science+Business Media, LLC 2009

Council UK (EC-UK) was established to coordinate and maintain the standards of the accreditation process. Thus, although there is, neither in France nor in UK, a formal obligation to register in order to practice as a professional engineer, in both countries the established standards provide a strong incentive for the accreditation of engineering degree programs (Augusti et al. 2007).

In addition to Great Britain and France, engineering program accreditation is an increasing practice in Europe, but, as described in several papers and reports (Augusti 2005, 2006) the situation varies considerably from country to country. For example, in Germany, up to a few years ago all higher education programs had to conform to strict (state or Federal) rules, which made accreditation superfluous. Bachelors and Masters programs were introduced in the late 1990s and are gradually replacing the old programs. Formal accreditation has been prescribed from the beginning for the Bachelors and Masters programs, and was later extended to all programs. A great number of German programs have been already accredited, especially in engineering.

In Portugal accreditation of engineering programs preceded the development of general Quality Assurance (QA) procedures. The order of Engineers established its accreditation procedure in 1994, well before the establishment of an overall QA system of higher education. In other countries (for example Italy, Lithuania, Romania, Switzerland, and several others) QA and program accreditation procedures are being introduced, although sometimes using different terminology, in the context of the so-called Bologna process, which is intended to establish the European Higher Education Area (EHEA).

It is fair to state that the quality of European engineering programs is generally quite high within the context of a global standard, and, on the whole, is continuously improving (thanks not only to QA practices but also to the continuous contacts and exchanges of good practices between engineering faculties). Such exchanges have been facilitated for several decades by international associations such as the Société Européenne pour la Formation des Ingénieurs (SEFI). More recently, EC-supported Thematic Networks on Engineering Education have emerged either for the whole of engineering or for specific branches.

## Motivation for a System of European Accreditation of Engineering Education

The variety of educational situations and degrees awarded in Europe makes transnational recognition of academic and professional qualifications rather difficult. The Bologna process is working toward the creation of a *transparent system* of easily readable and comparable degrees in the EHEA, but as far as professional accreditation and recognition are concerned, no generally accepted system or agreement exists on a continental scale. However, in engineering, several international agreements for mutual recognition of degrees and/or qualifications are active, for example, the Washington Accord (see section "The Global Context of EUR-ACE").

Notwithstanding the prestige of national systems and academic titles, this deficiency weakens the position of the European engineer in the global employment market.

The significance of this problem has been felt for quite some time. As early as 1994, the European Commission issued a communication on the possible synergies between the recognition of qualifications for academic and professional purposes (European Commission 1994). In 1998–1999 the Thematic Network, Higher Engineering Education for Europe (H3E) organized three European Workshops for Accreditation of Engineering Programs that led to the establishment in September 2000 of the European Standing Observatory for the Engineering Profession and Education (ESOEPE). It was quite natural for ESOEPE to respond to a March 2004 call for proposals by the European Commission (DG Education and Culture) stating that *the Commission supports the setting up and testing phase of transnational evaluation and accreditation* and *would welcome...proposals from subject specific professional organizations developing European Cooperation in Accreditation in fields like medicine or engineering*. The ensuing *EUR*opean *AC*redited *E*ngineer (EUR-ACE) project was launched in September 2004 and completed in March 2006.

## The EUR-ACE Project and the EUR-ACE Framework Standards

A main outcome of the EUR-ACE project (Augusti 2007) was a set of standards and procedures for accrediting engineering degree programs. A preliminary detailed survey of the standards used by project partners revealed striking similarities behind different façades, which made the derivation of a set of shared standards comparatively easy. Unlike the old national rules that prescribed inputs in term of subject areas and teaching loads, all the current Standards, and consequently the EUR-ACE Standards, define and require *learning outcomes*, i.e., what must be learned rather than how it is taught, an approach that has four direct advantages[1]:

1. It respects the many existing traditions and methods of engineering education (EE) in Europe.
2. It can accommodate developments and innovation in teaching methods and practices.
3. It encourages the sharing of good practice among the different traditions and methods.
4. It can accommodate the development of new branches of engineering.

[1]The US Accreditation Board for Engineering and Technology (ABET) was the first agency to shift in the late 1990s from a primarily input-based to a mainly outcomes- and performance-based accreditation with their so-called Engineering Criteria 2000 (*EC2000*). The ABET philosophy is dealt with in detail in the chapter "Quality Assurance in the Preparation of Technical Professionals: The ABET Perspective" by Peterson.

The definitive text of the EUR-ACE Framework Standards (European Network for Accreditation of Engineering Education 2008) was finalized after successive versions were commented on by the project partners and other stakeholders, both academic and nonacademic, and trial accreditations were run in a number of EHEA countries.

In accordance with the approach of the Bologna process, the EUR-ACE Standards distinguish between first and second cycle degrees (FCD, SCD) and identify 21 outputs for accredited FCD and 23 for SCD, grouped under six headings:

- Knowledge and understanding
- Engineering analysis
- Engineering design
- Investigations
- Engineering practice
- Transferable skills

The EUR-ACE Standards also contain guidelines and procedures for program assessment and program accreditation that include the assessment, among other requirements, of the human resources and facilities available for the program. They are consistent with the whole Bologna Process, and in particular with the Dublin Descriptors (Joint Quality Initiative 2004), the Framework for Qualifications of the EHEA [in short European Qualification Framework (EQF)] (Bologna Working Group on Qualifications Frameworks 2005), and the Standards and Guidelines for QA in the EHEA (European Association for Quality Assurance in Higher Education 2005). And they also take into account the EU Directive on the Recognition of Professional Qualifications (European Union 2005). Indeed, the EUR-ACE Framework Standards address the five generic qualification dimensions of the EQF on each level by specifying and expanding them with regard to engineering.

In order to be as flexible and comprehensive as possible, and not to exclude any European-compatible accreditation system, the EUR-ACE Standards encompass all engineering disciplines and profiles and distinguish only between FCD and SCD. However, the Standards are also applicable to the accreditation of programs leading directly to a degree equivalent to a SCD (conventionally termed Integrated Programs), that constitute an important part of European engineering education, and not only in the oldest continental Technical Universities Schools.

In some European countries, in addition to the distinction between FCD and SCD, engineering degrees are characterized by profiles; moreover, accreditation distinguishes between engineering branches (disciplines) in some countries, and not in others. The EUR-ACE Framework Standards can accommodate all these differences but they must be interpreted, and, if necessary, modified to reflect the specific demands of different branches, cycles, and profiles. However, they leave to Higher Education Institutes (HEIs) the freedom to formulate programs with an individual emphasis and character, including new and innovative programs, and to prescribe conditions for entry into each program.

A major difficulty in establishing program outcomes, and of differentiating between cycles, is that of specifying an absolute standard. This is particularly so in engineering because the standard must apply consistently to the many different and overlapping branches, and should also be applicable to new branches that will emerge as a result of continuing scientific and technical developments.

The EUR-ACE Framework expresses the standard to be achieved by FC and SC graduates in the three direct engineering requirements (Engineering Analysis, Engineering Design, and Investigations) by the phrase *consistent with their level of knowledge and understanding*, and this level is described using the concept of the forefront of the particular branch of engineering. For instance, in the requirement Knowledge and Understanding the relevant phrase is for First Cycle graduates, *coherent knowledge of their branch of engineering including some at the forefront of the branch* and for Second Cycle graduates *a critical awareness of the forefront of their branch.*

It would be extremely difficult, if not impossible, to obtain an agreed specification of the forefront for all engineering disciplines, and, even if it could be obtained, a fixed specification would inhibit innovation in program design and teaching methods. Nor would it be relevant or applicable to new and emerging technologies. The identification of the forefront of the branch is the responsibility of the members of the accrediting panel who are experts in that particular branch of engineering, while the body responsible for the final accreditation verdict will review and assess the rationale for their decision.

## The EUR-ACE Accreditation System and Its Implementation

The EUR-ACE Framework Standards do not intend to substitute for national standards, but to provide a common reference framework as the basis for the award of a common European quality label (the EUR-ACE label). Consequently, the EUR-ACE accreditation system was envisaged as based on a bottom-up approach involving the active participation of national accreditation agencies and leading to a multilateral mutual recognition agreement. No supranational Accreditation Board was proposed, i.e., accreditation will remain the task of national (or regional) agencies. This decentralized approach, now being implemented, appears to be rather novel in the world-wide panorama of program accreditation systems.

To implement the EUR-ACE system, ESOEPE has been transformed into the international not-for-profit association European Network for Accreditation of Engineering Education (ENAEE). ENAEE has registered the EUR-ACE trademark, and accredits (the term meta-accredits could be used) national agencies to add the EUR-ACE label to their accreditation.

ENAEE determined that six Accreditation Agencies in six different countries (namely, EC-UK; Engineers, Ireland; Order of Engineers, Portugal; RAEE, Russia; CTI, France; ASIIN, Germany) already fulfilled the requirements set by the Framework Standards and, in November 2006, accredited them to award the EUR-ACE label for a period of 2 years. Their accreditation, after a reassessment including site visits by multiagencies teams, was renewed for 2 more years starting from November 2008.

The six countries of this initial core of the EUR-ACE system cover a variety of educational, political, and social realities throughout Europe, such as to constitute a significant sample of the EHEA countries. Seventy-three programs obtained the EUR-ACE label in 2007, the first year of ENAEE operation, although only three agencies were ready to contribute. Approximately 200 labels have been awarded in 2008; many more are expected for 2009 and the following years.

## Spreading the EUR-ACE Accreditation System

Although the six countries constituting the initial core of the EUR-ACE system are a significant sample of the EHEA, their number is only about one-seventh of the total 46 EHEA countries. Therefore, ENAEE is now committed not only to strengthen the EUR-ACE system in these six countries, but also *to spread it* into other EHEA countries. Several paths are being followed to accomplish this aim, as illustrated by the following examples:

1. The Turkish Association for Evaluation and Accreditation of Engineering Programs (MÜDEK), promoted by the Turkish Engineering Deans' Council, started to accredit programs in 2003 and became an independent Association in 2007. MÜDEK applied to be EUR-ACE accredited and, after an accurate evaluation and site visits, was accredited in January 2009: thus, MÜDEK is the first example of another accreditation body specialized in Engineering programs joining the EUR-ACE system.
2. The Dutch–Flemish official Accreditation Organization NVAO (the only body legally authorized to accredit HE programs in the Netherlands and Flanders) has also applied in order to allow Dutch and Flemish engineering programs to be awarded the EUR-ACE label. This will be the first example of a *general* QA/ Accreditation Agency joining the EUR-ACE system pertaining to accreditation of engineering programs. Comparable Romanian and Lithuanian Accreditation Agencies (ARACIS and SKVC) are also currently in the pipeline to apply for EUR-ACE accreditation.
3. Some of the six core agencies already accredit engineering programs outside their own country; they have been authorized to award the EUR-ACE label also in such cases, and are starting to do so.
4. Individual HEIs from any EHEA country can apply, either to a specific Agency or ENAEE, to have their programs awarded the EUR-ACE label. This may be another way to start spreading the system into some countries. However, ENAEE plans a more systematic effort, especially in a number of countries where a specific interest has been expressed, for example, in Italy and Switzerland. This might possibly include the establishment of a new Engineering Accreditation Agency.
5. In principle, the EUR-ACE label may also be awarded outside the EHEA. Indeed, signals of interest for this possibility have already been sent to the ENAEE Headquarters. Of course, Path 4 above can already be followed, and similar systematic actions in countries outside the EHEA may well be planned in the future.

## The Global Context of EUR-ACE

Apart from the *European* context, EUR-ACE must confront the global scene, primarily in relation to the Washington Accord. This is an international agreement, started in 1989, among national accrediting bodies for engineering programs. Full members of the Washington Accord are agencies operating in USA (ABET), UK, Ireland, Canada, Australia, New Zealand, South Africa, Japan, Hong Kong China, Chinese Taipei, and Korea. Essentially, this agreement is among countries following a system of the Anglo-American type programs, with a first cycle (Bachelors) baccalaureate degree after 3 or 4 years of study and a second cycle (Masters) degree after 1 or 2 additional years.

The Washington Accord recognizes the substantial equivalency of programs accredited by the signatory bodies and recommends that graduates of programs accredited by any of them be recognized in the other countries. In this regard, the Washington Accord is analogous to the EUR-ACE system. However, the EUR-ACE system mutual recognition stems from a common quality label awarded by the participating agencies on the basis of shared standards and procedures (the EUR-ACE Framework Standards) while the Washington Accord relies on comparable accreditation procedures, independently applied by the participating agencies.

In most Washington Accord countries, one degree is the academic basis for entry into the engineering profession; therefore, the Accord recognizes only the Bachelors degree. However, this scheme is at present being questioned and there are pressures for the Washington Accord to move toward a two-tier system analogous to the Bologna/EUR-ACE scheme. Indeed, the EC-UK and Engineers Ireland (that are among the original signatories of the Washington Accord and also participate in the EUR-ACE systems) have accredited Masters degrees for a number of years. Beginning in 2009/2010, ABET will also allow accreditation of engineering programs provided by a HEIs at two levels (Bachelors and Masters).

The Washington Accord prescribes at least 4 years of study for an engineering Bachelors degree. In parallel, standards have been developed for 3- and 2-year program leading, respectively, to engineering technology degrees and engineering technicians' qualifications that are recognized in the so-called Sydney and Dublin Accords. The rigid and formal connection of outcomes with years of study and semantic definitions of technical professions in this three-accord (Washington–Sydney–Dublin) system causes difficulties in the mutual professional recognition for programs defined within the Bologna two-cycle[2] scheme, as well as for the academic recognition of such programs for graduates applying for admission to graduate studies.

Indeed, such problems should not exist in an outcomes approach. The assessment of certified-learning outcomes and gained competences should be independent from the ways of their achievement and the time it took. In this regard, the EUR-ACE Standards, consistent with the Bologna Process and the EQF, provide a

[2] The third cycle (doctoral studies) has been recently introduced in the Bologna process but is not yet considered in any accreditation scheme.

more flexible connection between outcomes and duration of study than do the Washington–Sydney–Dublin accords.

A comparison between the EUR-ACE and the Washington Accord requirements will be a crucial element in making the EUR-ACE label fully recognized globally, if for no other reason than that two members of the EUR-ACE core are also signatories of the Washington Accord. A comparative study is being promoted by ENAEE, and contacts have also been established with the International Engineering Alliance (IEA) that embraces the three Accords, in order to accomplish this aim.

## Conclusions

If coupled with rigorous QA rules, as it should always be, program accreditation assures that an educational program not only is of high academic standard, but also prepares graduates who are able to assume relevant roles in the job market. The participation of no-academic stakeholders in the process of setting standards and subsequent QA is a guarantee to this effect. An internationally recognized qualification like the EUR-ACE label, added to such an accreditation, will facilitate job mobility as well.

Engineering has always been in the forefront of discipline-specific accreditation, for example in France and the Anglo-Saxon countries, which has in many cases preceded the advancement of general QA procedures. Indeed, the engineering approach can be (and in some cases is) used as a model for other professional disciplines.

Discipline-specific accreditation is usually conferred on individual educational programs rather than departments or HEIs. However, this does not exclude and, on the contrary, is facilitated by an overall system of QA that authorizes only quality HEIs to deliver academic degrees.

When compared with the Washington–Sydney–Dublin Accord system it is fair to state that the EUR-ACE system is at the same time simpler and more flexible. This is the case since it does not create a rigid barrier between engineers and technologists, which is against the spirit of the Bologna Process, and in many languages even not understandable, but allows national differences and appropriate distinction between the cycles. Benchmarking the two systems will indeed be a major challenge for EUR-ACE. At the same time such an effort will be a test of the consistency and actual applicability of Dublin Descriptors (Joint Quality Initiative 2004), EQF (Bologna Working Group on Qualifications Frameworks 2005), and EU Directive on professional qualifications (European Union 2005).

But, apart from technical and operational difficulties inherent in creating a European scheme like the envisaged EUR-ACE system, a major difficulty lays certainly in the great differences between educational practices, legal provisions, and professional organizations across the different European countries. These are, however, the typical difficulties encountered in building a unified, but not homogenized, Europe.

The fact, that common Standards could be written and can be now implemented from Portugal to Russia, in continental and Anglo-Saxon countries, is a matter of great pride for us, initiators of EUR-ACE.

**Acknowledgments** The support by the DG "Education and Culture" of the EC through the projects EUR-ACE (2004–2006), EUR-ACE IMPLEMENTATION (2006–2008), PRO-EAST (2006–2007), and EUR-ACE SPREAD (2008–2010) is acknowledged. The author is also grateful to the many people, Institutions and Associations that have contributed to the EUR-ACE projects and are now collaborating with ENAEE, and in particular to the co-authors of Augusti et al. (2007, 2008) from which this paper draws heavily.

## References

Augusti, G. (2005). Trans-national recognition and accreditation of engineering educational programmes in Europe: Recent developments. *European Journal of Engineering Education, 30*(4), 417–422.

Augusti, G. (2006). Trans-national recognition and accreditation of engineering educational programmes in Europe: Perspectives in a global framework. *European Journal of Engineering Education, 31*(3), 249–260.

Augusti, G. (2007). Accreditation of engineering programmes: European perspectives and challenges in a global context. *European Journal of Engineering Education, 32*(3), 273–283.

Augusti, G., Freeston, I., Heitmann, G., & Martin, R. P. (2007). *Accreditation of engineering programmes as a tool to assure academic quality and relevance for the job market*. Presented at the 2nd European Quality Assurance Forum, Roma, November 2007. Accessed on October 26, 2008. Publications, presentations and other documents http://www.enaee.eu

Augusti, G., Freeston, I., Heitmann, G., & Martin, R. P. (2008). Accreditation and QA of engineering education in Europe: Setting up a pan-European system. *Implementing and Using Quality Assurance: Strategy and Practice: A Selection of Papers from the Second European Quality Assurance Forum, Roma, November 2007* (pp. 42–47); EUA.

Bologna Working Group on Qualifications Frameworks. (2005). *A framework for qualifications of the European higher education area*. Accessed on October 26, 2008. http://www.bologna-bergen 2005.no/Docs/00-Main_doc/050218_QF_EHEA.pdf

European Association for Quality Assurance in Higher Education (ENQA). (2005). *Standards and guidelines for quality assurance in the European higher education area*. Accessed on October 26, 2008. http://www.bologna-bergen2005.no/Docs/00-Main_doc/050221_ENQA_report.pdf

European Commission. (1994). *Communication on recognition of qualifications for academic and professional purposes*. Presented to the Council and the European Parliament, December, 1994.

European Network for Accreditation of Engineering Education (ENAEE). (2008). *EUR-ACE framework standards for the accreditation of engineering programmes*. As approved by the ENAEE Administrative Council on 5 November 2008. Accessed on October 26, 2008. http://www.enaee.eu/images/EUR-ACE_Framework%20Standards%20130109.pdf

European Union (EU). (2005). Directive 2005/36/EC of the European Parliament and of the Council on the Recognition of Professional Qualifications of 7 September 2005 on the Recognition of Professional Qualifications. *Official Journal of the European Union*, 30.9.2005, L 255/23-142. Accessed on October 26, 2008. http://eur-lex.europa.eu/LexUriServ/LexUriServ.do?uri=OJ:L:2005:255:0022:0142:EN:PDF

Joint Quality Initiative. (2004). *Shared Dublin descriptors for short cycle, first cycle, second cycle and third cycle awards*. Accessed on October 26, 2008. Joint Quality Initiative General Descriptors Webpage: http://www.jointquality.nl/

# Toward Consensus Global Standards for Quality Assurance of Engineering Programmes

**Hu Hanrahan**

**Abstract** This chapter contributes to the quest for a generic Global Model of Engineering Quality Assurance to guide future developments in engineering education Quality Assurance. Any Quality Assurance system requires two essential components: first, accreditation criteria, including exit-level outcomes, and second, policies and procedures for the programme evaluation process. This chapter describes a set of programme outcomes and level indicators, called Graduate Attributes (GA), developed by the International Engineering Alliance (IEA). The GA are related to a set of generic professional competency (PC) statements also developed by the IEA for the registration level. The GA and PC provide an understanding of the distinctive educational attributes and professional capability of the members of the engineering team: engineers, technologists and technicians. The chapter reviews the successes and limitations of the GA and how they could evolve.

## Introduction

In the modern world, people, societies and economies are dependent on infrastructure, services and the availability of goods. Meeting these needs calls for a wide range of activities, including the exploitation of natural resources, construction, manufacture, energy supply, communication, transportation, utilities as well as the control of complex processes and effective use of information. These activities require engineering skills for technically sound, safe, economical and sustainable execution. Achievement of effective and sustainable solutions with risks mitigated to an acceptable level depends on the competence of engineering practitioners. The development of competent engineering practitioners includes an educational requirement followed by training and experience before professional recognition. This chapter examines the essentials of engineering competence, the role that

H. Hanrahan (✉)
School of Electrical and Information Engineering, University of the Witwatersrand, Johannesburg, Private Bag 3, WITS 2050, South Africa
e-mail: h.hanrahan@ee.wits.ac.za

A.S. Patil and P.J. Gray (eds.), *Engineering Education Quality Assurance: A Global Perspective*,
DOI 10.1007/978-1-4419-0555-0_4, 

engineering education plays in developing engineering competence and how consensus standards for the education of engineering practitioners have developed.

As economic activity grows more global, engineering is practised across national boundaries and engineering professionals are internationally mobile. The quality of engineering practice is therefore important not only within but also across national boundaries. The quality of engineering education programmes can no longer be a national concern only. Various arrangements have therefore come into existence for the recognition of engineering qualifications on a regional or global basis. One such arrangement for the mutual recognition of engineering educational programmes and to facilitate mobility of professionals is the International Engineering Alliance (IEA). The IEA has three constituent educational accords: the Washington Accord (WA), the Sydney Accord (SA) and the Dublin Accord (DA), for engineers, engineering technologists and engineering technicians, respectively. Two agreements seek to aid the mobility of registered engineers and technologists, namely the Engineers Mobility Forum (EMF) and Engineering Technologist Mobility Forum (ETMF).

An objective of this book is to present a generic Global Model of Engineering Quality Assurance based on the lessons learned in existing arrangements. Any model for the Quality Assurance of engineering education programmes has two important facets: the standards that graduates must meet and the best-practice processes for quality assuring programmes. This chapter addresses the standards aspects, together with insights into their development, usage and ongoing evolution. This chapter contributes to an understanding of widely accepted standards for engineering education programmes and the consensus on the desired attributes of graduates of education programmes for engineers, engineering technologists and engineering technicians developed by signatories of the three educational accords.

## Setting the Context for Engineering Education Quality Assurance

This chapter is concerned with the standards that would be used in the design and Quality Assurance of engineering education programmes. These standards must be contextualised in the lifecycle of an engineering practitioner. The development of an engineering practitioner has the principal stages shown in Fig. 1.

An educational qualification is first obtained. The graduate then enters a programme of training and gains experience to develop the competency required for registration. The candidate then applies to be registered and undergoes competency assessment prescribed by the registering authority. Important elements of engineering professional development and their relationships are shown in Fig. 1.

At the educational stage, three interacting elements are involved in the provision of quality education. First, the standards set by the accrediting body define the required attributes of the graduate as well as knowledge requirements. Second, the engineering programme is designed with educational objectives and assessable outcomes that are evidence that the programme meets its objectives.

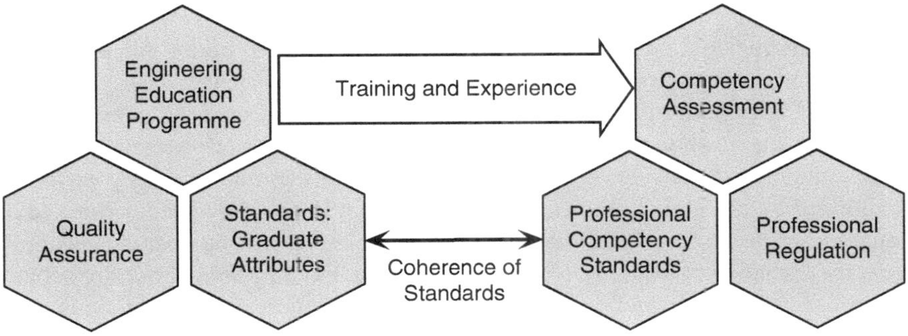


**Fig. 1** Graduate attributes and professional competencies in context

The outcomes set by the provider must include those defined in the standard. Third, the educational programme is subject to an external Quality Assurance process that evaluates the achievement of the programme against the standard and other criteria such as programme structure, the quality of teaching and learning and the resourcing and sustainability of the programme.

Having attained an engineering qualification that is accredited or recognised under a mutual recognition agreement or evaluated on an individual basis as meeting educational requirements, a person enters a training programme, working with and under the supervision of qualified professionals. The trainee typically starts by assisting the established professionals and, as proficiency grows, is allowed to work under diminishing detailed supervision and to assume increased responsibility. The objective is to develop the competency that must be demonstrated for registration.

At the registration stage, there are also three interlocking elements. First, a professional body is responsible for admitting persons to registration and regulating their practice. Second, the level of competence required for registration is defined in competency standards. The standards to be met by the educational programme must be coherent with those for the professional level: they provide the entry and exit boundary conditions for the process of training and experience. Third, assessment of competence is performed by the professional body.

A third possible milestone for engineers and engineering technologists in jurisdictions that are signatories to engineering mobility agreements is being admitted to an international register. Additional experience is required beyond professional registration and a position of responsibility must have been held.

## International Engineering Education Accords

Three educational accords span the engineering team: engineers, engineering technologists and engineering technicians. The oldest of the accords is the Washington Accord (WA), signed in 1989. This agreement provides for mutual recognition of the qualifications accredited as meeting educational requirements toward professional engineering registration among its signatories. The Washington Accord[1] came about because six national engineering bodies felt that, on the basis of mutual knowledge, they had sufficient confidence in each others' accreditation criteria and procedures that each would be willing to accept the accreditation decisions of the other signatories. Graduates of a programme accredited by one signatory would be recognised by the others. Subsequently, six further bodies have been admitted as signatories. The current signatories are listed in Table 1. As the number of signatories grew, the operation of the accord became formalised. Existing signatories are subject to periodic verification of their standards and procedures. Aspiring signatories spend time in provisional status before demonstrating that their standards and procedures warrant admission as a signatory.

The Sydney Accord (SA) for the mutual recognition of engineering technologist education programmes accredited by signatories was signed in 2001 by seven

**Table 1** Signatories of educational accords and mobility fora as of June 2007

| Body and jurisdiction | WA | SA | DA | EMF | ETMF |
|---|---|---|---|---|---|
| ABET, Inc. (USA) | • | | | | |
| Accreditation Board for Engineering Education of Korea (ABEEK) | • | | | | |
| Canadian Council for Professional Engineers (CCPE) | • | | | • | |
| Canadian Council for Technicians and Technologists (CCTT) | | • | • | | • |
| Engineers Australia | • | • | | • | |
| Engineering Council of South Africa (ECSA) | • | • | • | • | • |
| Engineering Council United Kingdom (EC UK) | • | • | • | • | • |
| Hong Kong Institution of Engineers (HKIE) | • | • | | • | • |
| Institution of Engineers, Ireland (Engineers Ireland) | • | • | • | • | • |
| Institution of Engineers Malaysia (IEM) | | | | • | |
| Institute of Engineering Education Taiwan (IEET) | • | | | | |
| Institution of Professional Engineers New Zealand (IPENZ) | • | • | • | • | • |
| Institution of Engineers, Singapore (IES) | • | | | • | |
| Institution of Engineers Sri Lanka (IESL) | | | | • | |
| Institution of Professional Engineers Japan (IPEJ) | | | | • | |
| Japan Accreditation Board for Engineering Education (JABEE) | • | | | | |
| Korean Professional Engineers Association (KPEA) | | | | • | |
| United States Council for International Engineering Practice (USCEIP) | | | | • | |

Note: Washington Accord (WA); Sydney Accord (SA); Dublin Accord (DA); and Professional Level Mobility Agreements for Engineering (EMF) and for Engineering Technologists (ETMF)

[1] Washington Accords (1989) at http://www.ieagreements.org/Washington-Accord/

accrediting bodies. The Sydney Accord operates similarly to the Washington Accord and The Dublin Accord (DA) was signed in 2002 for mutual recognition of engineering technician education.[2] Because engineering technician education systems differ significantly from country to country and because national rather than professional accreditation systems may be in operation, this accord provides for the recognition of exemplifying academic qualification types rather than individual education provider's programmes. The SA and DA signatories are shown in Table 1. Also shown are the current signatories to the professional level mobility agreements for engineers (EMF) and engineering technologists (ETMF).

The accords are premised on the importance of education as a foundation for practice at the professional level. Signatories to an accord recognise that the criteria, policies and procedures of other signatories are comparable and agree that their accreditation decisions are mutually acceptable. This confidence is based on the verification of criteria, policies and procedures prior to admission as a signatory and periodic monitoring of existing signatories.

The acceptance of accreditation decisions of other signatories is based on substantial equivalence of programmes (WA and SA) and of the exemplifying educational qualification types (DA). Substantial equivalence applied to educational programmes means that two programmes, while not meeting a single set of criteria in detail, are both acceptable as an education base that prepares their respective graduates to enter training and experience toward registration in a jurisdiction. Substantial equivalence recognises that, within limits, education requirements can be satisfied in different ways and graduates could, subject to individual adaptations, train toward professional registration on an equal footing. Substantial equivalence is in some ways analogous to thermal equilibrium. Imitating the Zeroth Law of Thermodynamics: If two educational quality systems are each substantially equivalent to a third, they are also substantially equivalent to each other.

Prescriptive standards are therefore not essential to the mutual recognition process; however, an objective base for substantial equivalence is helpful and that base is provided by the Graduate Attributes (GA) described below.

## Developments in Engineering Education and Accreditation

While the international educational accords developed, engineering education was evolving. A significant paradigm shift took place in engineering education and accreditation practice during the 1990s. At the end of the 1980s, the prevailing engineering education paradigm emphasised curriculum structure, content and the technical depth achieved. From 2000, the emphasis in curriculum design and

[2] Sydney Accords (2001) at http://www.ieagreements.org/sydney/ and Dublin Accords (2002) at http://www.ieagreements.org/Dublin/

accreditation fell on the attributes the graduate is expected to possess, without losing sight of content but not seeking depth in all the areas. The transition from depth to outcomes was characterised by debates about broadening the curriculum.

In the phase focusing on depth of knowledge, the curriculum was premised on the assumption that the engineering graduate would slip seamlessly into a job where the technical knowledge of the curriculum would be applied. Engineer education courses provided a base of fundamentals and took students to technical depth in the major area of the curriculum. In some countries, there was a requirement for humanities and social sciences in the curriculum as a general educational requirement. Toward the end of this period, curriculum reform was driven by the need to make the curriculum more interesting and attractive to students (Director et al. 1995; Miksad et al. 1996). The main changes were introducing students to engineering topics and design at an early stage. Design emerged as a central theme in curricula and also as an emphasis in accreditation criteria. There was also recognition that most engineering disciplines had grown to the extent that graduates could not be expected to have deep knowledge in all the areas of their discipline.

The transition to the outcomes-oriented paradigm was driven by the realisation that graduates do not progress from academic study in a technical field to a job for which the curriculum's specific knowledge and skills are sufficient. Rather, technology and engineering applications are changing as is the environment in which engineering is practised. Engineering graduates progress rapidly to the supervision of people, the management of projects, control of finances and dealing with risk. They must cope with ever changing knowledge, technology, applications and environment (van Valkenberg 1990). The need was recognised for graduates to communicate well, to work with persons in other disciplines, to continue learning and to deal with the impacts of engineering activity. Engineering at the professional level not only rely on natural sciences and technical prowess but also needed to engage with the needs of people, societies and economies and, increasingly, the physical environment (Wenk 1997). These requirements were encapsulated in the notion of breadth of engineering education. Debates on breadth vs. depth and broadening the curriculum abounded and the social and environmental dimension of engineering moved to the foreground (Wenk 1997).

Turning the broadening requirements into curricula and accreditation criteria raised challenges. Desiderata such as "graduates should be strongly analytical, should have practical ingenuity, creativity, well developed communication skills, leadership, should understand ethics and professionalism" needed to be cast in a form that is assessable, does not have to be changed with rapid changes in technology and also applies to all engineering disciplines. The introduction of broadening considerations should not compromise the basic scientific and technical aspects of the curriculum. A number of forward looking studies, for example (Institution of Engineers Australia 1996; Lang et al. 1999), provided a way of implementing broadening and engineering requirements by identifying a set of generic outcomes or GA. Many of these developments reflected a better understanding and expression of what engineering educators, quality assurers and professional bodies already understood intuitively. Various accrediting bodies developed programme criteria that capture the requirements for fundamental and specialist knowledge as well as deepening requirements in a generic form.

The new accreditation criteria did not diminish the importance of knowledge as a GA. The emphasis on mathematics, physical science and engineering fundamentals as was reinforced. Specialist study, that is achieving technical depth in an area, was seen as important in the curriculum. However, there was a sober realisation that, if justice is to be done to the fundamentals, it is impossible to accommodate a large body of specialist knowledge in an undergraduate curriculum. Realistically, it is not possible in the undergraduate curriculum to foresee and provide all the specialist knowledge a graduate will need. Rather, the undergraduate programme must develop a base of fundamentals and an ability to continue learning. The ability to learn independently became a programme outcome. Postgraduate programmes were seen as providing the opportunity for further formal specialist study.

By 2000, several Washington Accord signatories had completed or were in the process of defining outcomes-based accreditation criteria. They also adjusted their accreditation evaluation processes to accommodate the focus on assessable outcomes that encapsulate the desired attributes of graduates.[3] While all of these initiatives defined the attributes of graduates of engineer programmes, several also formulated outcomes for engineering technologist and engineering technician education. It was therefore natural in the forum created by the educational accords for signatories to ask whether a consensus set of GA could be formulated for the whole engineering team.

## Toward and Objective Basis for Substantial Equivalence: The Graduate Attributes

In the late 1990s, movement occurred in many of the WA signatories toward understanding the outcomes of engineering education and incorporating these into their accreditation criteria. Balanced accreditation criteria emerged including curriculum structure and broad content profile, the required attributes of the graduate, the quality of the teaching and learning and the resources to sustain the programme. *Graduate attributes* (GA) are statements of what the graduate

[3]See for example: ABET, Criteria For Accrediting Engineering Programs (2005), Available http://www.abet.org; Engineering Council UK, The Accreditation of Higher Education Programmes, UK Standard for Professional Engineering Competence, 2004, Available http://www.engc.org.uk; Engineers Australia, National Generic Competency Standards – Stage 1 Competency Standards for Professional Engineers, Document P05, 2005; Engineers Ireland, Accreditation Criteria for Engineering Programmes, 2003, Available http://www.engineersireland.ie/; IPENZ , Requirements for Initial Academic Education for Professional Engineers, December 2003, http://www.ipenz.org.nz/ipenz/forms/pdfs/

should be able to do that would be assessed by the education provider. GA identify elements that are assessable and that give confidence that a programme is meeting generally accepted objectives.

In 2001, the WA signatories observed that, in the process of affirming substantial equivalence among their fellow signatories, they had come to acknowledge generally accepted, globally relevant attributes which graduates from accredited engineering programmes are expected to possess. The signatories accepted a proposal that a study be undertaken to develop descriptions of the competence that a graduate engineer from a Washington Accord accredited degree could reasonably be expected to exhibit.

Many educational accord signatories who are also registering bodies were also looking to develop statements that would describe the competency required for registration as an engineer, an engineering technologist or an engineering technician. The EMF had been established to facilitate the mobility of professional engineers through the establishment by each signatory of an International Register. The criteria for admission to the register are based on education, registration, additional experience and responsibility, representing competence beyond that required for registration. The International Register of Professional Engineers, Constitution (Engineers Mobility Forum 2007) envisaged that signatories could, over time, develop an alternative route to the International Register via competency assessment, using work-derived evidence against standards. Signatories of the EMF therefore decided to prepare consensus statements of *professional competency* (PC). Because of the need to make GA coherent with competence at the professional level, the EMF and WA, a joint development process was agreed. Neither competency description was intended at that time to form part of either agreement, but to provide information that would inform and enable evolving practices to be shared.

The initiative thus far was concerned solely with engineers. Several signatories were engaging with educational outcomes and PCs for engineering technologists and engineering technicians. The ETMF had been established along the lines of the EMF. In 2003, the Institution of Professional Engineers New Zealand (IPENZ) made a substantial input into signatories' deliberations that defined a consistent approach to engineers, engineering technologists and engineering technicians at both the educational and professional levels. This approach is the basis of the GA and PC described below. The initiative to develop GA and PC statements was extended to include technologists and technicians.

An International Engineering Workshop was held in 2004 by the educational accords and mobility fora. By that time, numerous signatories had published educational outcomes and competency statements and it was possible through analysis to identify a significant number of common attributes; in fact, few outcomes were unique to individual signatories. The Workshop defined GA and PC for engineers, engineering technologists and engineering technicians. Minor changes followed circulation of the statements to the Signatories. The GA and PC were adopted by the educational accords and mobility fora in June 2005 (International Engineering Alliance 2005). In 2007, the educational accords incorporated the

GA into their operating rules and procedures as exemplars of graduate competency (International Engineering Alliance, 2007).

Before describing the method used to define the GA and the actual statements, we present the statements of their agreed purpose and limitations. The GA are a set of individually assessable outcomes that are components of the graduate's potential competence, that is, exemplars of the attributes expected of a graduate from an accredited programme. The GA are not intended to be an international standard to which all accredited programmes should conform. Rather, the GA are intended to assist national bodies to develop outcomes-based accreditation criteria and to guide bodies developing their accreditation systems with a view to seeking signatory status.

The GA, having been defined for engineer, technologist and technician education programmes, are useful in defining the commonality and difference between the outcomes of the different types of programmes and between graduates of the three types. Their distinctive roles in forming an engineering team are made plain.

## *Method of Defining Graduate Attributes and Professional Competencies*

The GA cover the exemplifying qualifications for engineers, engineering technologists and engineering technicians. The process of developing GA was aided significantly by the ongoing efforts by signatories to shift accreditation criteria from content and educational process to outcomes. Signatories had already addressed questions such as the following. What are the key attributes? How do they vary from engineer to engineering technologist to engineering technician? What level of performance is required? How to deal with the diversity of engineering while keeping the definition of attributes finite?

The approach was based on identifying common features of graduate competence for engineers, engineering technologists and engineering technicians and differentiating only where necessary. Attribute definitions were kept generic; capturing what is common to all engineering disciplines.

## *Identifying the Outcomes*

A consensus had emerged that graduate engineers, engineering technologists and engineering technicians entering employment have common functions, for example problem solving, the use of knowledge and communication. The expected nature and level of performance is, however, different across the three types of graduate. The GA cover 13 areas (International Engineering Alliance 2005) as follows:

1. *Academic education*. An indication of the normal extent of higher education required for the qualification accredited for the category of registration. This is a measure of the body of knowledge rather than an assessable competency such as 2, 3 and 4 below that deal with the application of knowledge.
2. *Knowledge of engineering sciences*. What the graduate is expected to do with engineering science knowledge, supported by mathematical and basic science knowledge.
3. *Problem analysis*. The first component of a problem solving ability, the core activity of engineering.
4. *Design/development of solutions*. The solution phase of problem solving where solutions must be synthesised, evaluated and refined meet technical, economic, social and environmental criteria.
5. *Investigation, experimentation and analysis*. Reflecting the ability to explore problematic situations, perform research, obtain answers by observation and measurement and analyse results obtained.
6. *Modern tool, techniques and other resource usage*. Reflecting the fact that engineering activity is supported by many aids: computational, procedural and measurement.
7. *Individual and team work*. Reflecting the reality that graduates must be effective workers in their own right and as members of teams.
8. *Communication*. Written and oral, with both technical and non-technical audiences.
9. *The engineer and society*. Understanding of issues arising in engineering activities.
10. *Ethics and ethical behaviour*. Reflecting the need for graduates to understand ethics and to act ethically.
11. *Environment impacts and sustainability*. Reflecting the need for engineers to predict and detect the impact of engineering activity on the environment and to incorporate sustainability consideration into their work.
12. *Project management, finance, risk and change*. The essential management actions in support of engineering activity.
13. *Life-long learning*. Reflecting the need for a demonstrated ability to perform learning at the requisite level of independence.

Performance areas 2–13 capture both the desired technical ability of the graduates and encapsulate the broadening requirements.

Within each performance area, a statement of the attribute is formulated for each type of graduate. The GA are formulated to be individually assessable in the sense that discrete assessment tasks can be set for each element in the educational environment. This does not rule out holistic assessment, with more than one outcome being assessed in a single task. Each outcome statement has a common stem. For example, the design attribute at the graduate level uses a common wording, with ranging information inserted wherever a square bracket is shown, as given below:

Design solutions for [level indicator/type of] problems and [perform action] systems, components or processes that meet specified needs with appropriate consideration for public health and safety, cultural, societal, and environmental considerations.

The variable part of the outcome statement defines the differentiating characteristic, the different ways that engineers, engineering technologists and engineering technicians deliver the basic outcome. For the graduate level design outcome, the differentiating characteristic is the *breadth and uniqueness of engineering problems*, that is, *the extent to which problems are original and to which solutions have previously been identified or codified*. The resulting attribute statements for the three types of graduates are shown in Table 2.

The three statements in Table 2 are distinguished by an indicator of the type of problem solving involved (complex, well-defined and broadly defined) and the role in the design process (designing, contributing to design and assisting with the design process). At the outset, we stress that the terms complex engineering problems, broadly defined engineering technology problems and well-defined technical problems provide a convenient shorthand for the full descriptions given in International Engineering Meetings 2005[4] and shown in Table 3 (International Engineering Alliance 2005).

The notion of a range of problem solving applies to both GA and PCs. In addition, the types of challenge that engineers, engineering technologists and engineering technicians are expected to meet at the professional level are described by the notions of complex engineering activities, broadly defined engineering activities and well-defined engineering activities. The professional is expected to produce specified outcomes within the type of engineering activities. The full set GA and PCs are posted on the International Engineering Agreements Web site.[5]

**Table 2** Stage 1 engineering design attribute

| Engineer | Engineering technologist | Engineering technician |
|---|---|---|
| Design solutions for complex engineering problems and design systems, components or processes that meet specified needs with appropriate consideration for public health and safety, cultural, societal and environmental considerations | Design solutions for broadly defined engineering technology problems and contribute to the design of systems, components or processes to meet specified needs with appropriate consideration for public health and safety, cultural, societal and environmental considerations | Design solutions for well-defined technical problems and assist with the design of systems, components or processes to meet specified needs with appropriate consideration for public health and safety, cultural, societal and environmental considerations |

[4] See International Engineering Alliance, International Engineering Meetings (IEM) at http://www.washingtonaccord.org/IEM.cfm

[5] International Engineering Agreements Web site: http://www.ieagreements.org/

**Table 3** Problem-solving classification (International Engineering Alliance 2005)

| *Complex engineering problems* cannot be resolved without in-depth engineering knowledge and have some or all of the characteristics: | *Broadly defined engineering problems* have some or all of the following characteristics: | *Well-defined engineering problems* have some or all of the following characteristics: |
|---|---|---|
| Involve wide-ranging or conflicting technical, engineering and other issues | Involve a variety of factors which may impose conflicting constraints | Involve several issues, but with few of these exerting conflicting constraints |
| Have no obvious solution and require abstract thinking, originality in analysis to formulate suitable models | Can be solved by application of well-proven analysis techniques | Can be solved in standardised ways |
| Requires in-depth knowledge that allows fundamentals-based first principles analytical approach | Requires knowledge of principles and applied procedures or methodologies | Can be resolved using limited theoretical knowledge but normally requires extensive practical knowledge |
| Involve infrequently encountered issues | Belong to families of familiar problems which are solved in well-accepted ways | Are frequently encountered and thus familiar to most practitioners in the practice area |
| Are outside problems encompassed by standards and codes of practice for professional engineering | May be partially outside those encompassed by standards or codes of practice | Are encompassed by standards and/or documented codes of practice |
| Involve diverse groups of stakeholders with widely varying needs | Involve several groups of stakeholders with differing and occasionally conflicting needs | Involve a limited range of stakeholders with differing needs |
| Have significant consequences in a range of contexts | Have consequences which are important locally, but may extend more widely | Have consequences which are locally important and not far reaching |
| Are high level problems possibly including many component parts or sub-problems | Are parts of, or systems within complex engineering problems | Are discrete components of engineering systems |

## *Contextual Interpretation*

The generic statements of the GA are intended for the application in all engineering disciplines and require interpretation into different contexts: categories of professional, engineering disciplines or industry sectors. This interpretation is performed by peers, that is, persons who are qualified and experienced in the relevant category and discipline. This practice is well established in accreditation and professional registrations systems.

## The Professional Competencies

The PCs are stated for each of the three categories using 13 elements listed in International Engineering Alliance (2005) Graduate Attributes and Professional Competencies. The PC Elements are conveniently grouped as follows:

(a) *Engineering knowledge*. Good practice in the professional category is underpinned by a sound knowledge base having several components. The fundamentals from the education phase remain important as they underpin specialist knowledge. Specialist knowledge in the practice area is essential, both universally applicable and particular to the local conditions. The latter may be legal and regulatory but also relate to local conditions such as available materials and climatic conditions.

(b) *Problem analysis, solution and evaluation*. Problem solving is the core process that an engineering practitioner performs on a regular basis. Problems that the practitioner must be capable of solving are specified by the range indicator. In addition, these problems must be solved within activities at appropriate levels of demand for the three categories.

(c) *Impacts of engineering activity*. Here there are two dimensions. First, all legal and regulatory requirements relating to the work must be satisfied. Second, the effects of engineering solutions on people, communities and the environment must be foreseen and adverse effects mitigated to an acceptable level.

(d) *Managing engineering activity*. This group reflects an aspect of the broadening requirements since engineering work is not purely technical but requires the engineer to make things happen and to control resources.

(e) *Acting ethically, applying judgment and being responsible*. This group is concerned with three very personal attributes: first, being able to recognise and deal with ethical problems, second, being able to make decisions in the absence of full information and, third being able to make decisions in a responsible manner and be willing to be accountable for those decisions.

(f) *Life-long learning*. At the educational level, the requirement is to develop and demonstrate an independent learning ability. At the professional level, the key competence is to actually maintain and extend competence.

At the professional level, evidence of competence will come from workplace activity.[6] As the organisation of this work is dictated by the requirements of the employer, it is unlikely to be decomposed in a way that individual outcomes can be assessed. Competence, in any event, is the ability to perform the required job functions. For these two reasons, assessment of competence must be holistic. Sets of evidence derived from the workplace demonstrate performance against several outcomes. For example, in an engineering design, the engineer must identify problems, synthesise solutions and evaluate results, taking both performance requirements and impacts into account, while complying with legal requirements.

[6] It is recognised that professional practice examinations are used in some jurisdictions.

**Table 4** Describing the level of engineering activity (International Engineering Alliance 2005)

| *Complex engineering activities* have some or all of the following characteristics: | *Broadly defined engineering activities* have some or all of the following characteristics: | *Well-defined engineering activities* have some or all of the following characteristics: |
|---|---|---|
| Involve the use of diverse resources (and for this purpose resources includes people, money, equipment, materials, information and technologies) | Involve a variety of resources (and for this purposes resources includes people, money, equipment, materials, information and technologies) | Involve a limited range of resources (and for this purpose resources includes people, money, equipment, materials, information and technologies) |
| Require resolution of significant problems arising from interactions between wide-ranging or conflicting technical, engineering or other issues | Require resolution of occasional interactions between technical, engineering and other issues, of which few are conflicting | Require resolution of interactions between limited technical and engineering issues with little or no impact of wider issues |
| Involve creative use of knowledge of engineering principles in novel ways | Involve the use of new materials, techniques or processes in innovative ways | Involve the use of existing materials, techniques or processes in new ways |
| Have significant consequences in a range of contexts | Have consequences that are most important locally, but may extend more widely | Have consequences that are locally important and not far reaching |
| Can extend beyond previous experiences by applying principles-based approaches | Require a knowledge of normal operating procedures and processes | Require a knowledge of practical procedures and practices for widely applied operations and processes |

Definition of the PCs follows the same logic as for the GA. Thirteen performance elements are defined. A common stem is used across the engineer, engineering technologist and engineering technician forms. The level of performance is distinguished using the gradation of problem solving already defined as well as levels of engineering activity. Three sets of descriptors are defined as shown in Table 4 and are labelled *complex engineering activities, broadly defined engineering activities and well-defined engineering activities* for the three professional categories.

## Progression from Graduate Attributes to Professional Competencies to International Register

The main learning process between graduation and professional registration is through working with competent engineering personnel. The trainee is under the direct or indirect supervision of an engineering professional while a mentor guides the trainee's professional development. The trainee performs engineering work of adequate variety and increasing demand and responsibility. The candidate would first *assist* with engineering work, doing defined tasks under close supervision. The candidate progresses to making contributions to engineering work, both individually and as a team member. By the end of the training period, the candidate must perform individually and as a team member at the level required for registration, thereby providing evidence of competency against the standards. Over time, the emphasis on *training*, that is, learning through inputs of others, gives way to learning by doing engineering work and reflecting on observations and achievements, that is *experience*.

The GA define the initial capability of the trainee. Training and experience must develop the trainee's capability to the level defined by the PCs. Designers of training programmes and mentors require the understanding of changes in competence outlined next. We use the classification of PC into groups (a)–(f) and the engineer category by way of illustration.

GA-2 (with GA-1 indicating the typifying duration of study) requires that the graduate has a systematic, theory-based formulation of engineering fundamentals of the discipline, together with engineering specialist knowledge typical of the discipline. This knowledge is built on a base of mathematics and natural sciences. The graduate also comprehends the role of engineering in society, ethics and the impacts of engineering activity: economic, social, cultural, environmental and sustainability. Group (a) of the professional competencies indicates both universal and jurisdiction-specific knowledge. This includes deepened, contextualised engineering knowledge in the same or different specialist areas, together with comprehensive knowledge that supports design, investigation and solution synthesis. Practical insight is required into the ethical, economic, social, cultural and environmental impacts and sustainability and methods for addressing these.

At the graduate level, GA 3–6 relate to problem solving within the context of an academic programme. Problem solving ability at the graduate level is at the complex level defined in Table 3. At the professional level, group (b) of the PC defines the required problem analysis, solution design and evaluation competencies. Problems that the professional must be capable of addressing are also at the complex engineering problem level but must be demonstrated in an actual work environment, characterised by the description of *complex engineering activities* defined in Table 4. The work used to demonstrate competency may take on many forms including conventional design and investigations.

At the graduate level, GA-11 requires that the graduate has an understanding of social and environmental issues, demonstrated in an academic context. At the professional level, group (c) of the PC refers to impacts of actual engineering activity on society, the environment and sustainability that must be identified and mitigated to an acceptable level. Similarly, all legal and regulatory conditions in the jurisdiction must be satisfied. The professional must be able to fulfil these requirements while performing complex engineering activities.

GA outcomes 7, 8 and 12 define the personal and management-related abilities expected of the graduate: effective team and individual work in a multi-disciplinary (academic) setting (GA-7); communicative competence with an audience at large but demonstrated to academic assessors (GA-8) and knowledge of management principles (GA-12). At the professional level, group (d) of the PC encompasses communication and the management abilities needed to effectively execute engineering activity. The ability to manage complex engineering activities must be demonstrated. The engineer must therefore deal with the types of factors listed in Table 4.

At the educational level, GA 9 and 10 require that professional and ethical principles must be understood and issues handled in a simulated environment. In group (e) of the PC, the professional must be able to apply judgement in decision making, act responsibility and be accountable for complex engineering activities.

Finally, GA 13 is concerned with the development of an ability and inclination to learn independently as a foundation for ongoing professional development. Group (f) of the PC reinforces this continuing need.

## Evolution of the Graduate Attributes

The GA arose first out of an exploration of the attributes of a Washington Accord graduate. They were broadened to create an understanding of the attributes technologists and technician graduates and to express an understanding of the distinctive roles within the engineering team. The GA then became exemplars of graduate competency in the Rules and Procedures for the educational accords (International Engineering Alliance, 2007). There has been a clear understanding that they do *not* constitute an *international standard* to which all signatories' domestic accreditation criteria must comply. A high degree of congruence has nevertheless emerged between programme outcomes defined by signatories for their accreditation processes and the GA. This is not surprising: the development of GA has been assisted by and has assisted national bodies developing their outcomes-oriented criteria.

Signatories reported in 2007 on their experience with the GA. Several signatories recounted using the GA as a reference when formulating their own criteria. Some reported that the GA have proved useful in understanding and explaining the differences in expectations between programmes in engineering and engineering technology.

The GA continue to evolve with issues such as the following being addressed. What is the fundamental purpose of engineering education and how do the GA relate to this purpose? Do changes that are taking place in national education systems and professional formation requirements prompt changes to the GA? Issues such as these are under review by the IEA and readers should refer to the current version of the GA and PC posted on the IEA Web site.[7]

---

[7]http://www.ieagreements.org

## Lessons Learned

The GA and PCs, while enjoying significant consensus will continue to evolve. Perhaps the most sobering lesson from the development process is the difficulty of distilling the individual and collective understanding of outcomes and competence into a statement that persons who have not been party to the process can understand: what engineers understand intuitively is difficult to capture on the page.

A second lesson is that an understanding of the distinctive roles, competencies and educational requirements of engineers, technologists and technicians is beneficial to the design and accreditation of education programmes, to the effective employment of the different team members and to regulation of practice.

A third lesson emerges from the question: why have the IEM signatories succeeded in developing GA and PC that enjoy widespread support? The participants would like to think that success flows from the fact that the accords are associations of national authorities that have banded together voluntarily because they see individual and mutual benefits. The Washington, Sydney and Dublin Accords represent a step towards globally consistent Quality Assurance of engineering education programmes. The experience of the signatories suggests that a global model of Quality Assurance should be based on a voluntary rather than enforced participation.

Finally, there was remarkable consistency among standards set by various registering bodies as the minimum each considers necessary in each category of registration for the professional to function effectively, economically, safely and in an environmentally sound manner. As the international norms are a distillation of the national baselines, they reflect generally acceptable minimum standards for technicians, technologists and engineers; "international" does not mean a higher standard.

**Acknowledgements** The author's grateful thanks go to George Peterson (author of chapter "Quality Assurance in the Preparation of Technical Professionals: The ABET Perspective") who first proposed defining graduate attributes; Andrew Cleland, who took up the challenge and Basil Wakelin who crystallised the unified approach to Graduate Attributes and Professional Competencies. Many colleagues from IEA signatories have contributed regularly to the ongoing development the GA and PC including Kate Aberle, Jim Birch, Alan Bradley and Deborah Wolfe. The kind permission of the IEA Secretariat to reproduce material in Tables 3–5 (the Appendix) is acknowledged and greatly appreciated.

## Appendix

See Table 5.

**Table 5** Graduate attributes (International Engineering Alliance 2005)

| Characteristic | …for Washington Accord graduate | …for Sydney Accord graduate | …for Dublin Accord graduate |
|---|---|---|---|
| 1. Academic education: Educational depth and breadth | Completion of an accredited programme of study typified by 4 years or more of post-secondary study | Completion of an accredited programme of study typified by 3 years or more of post-secondary study | Completion of an accredited programme of study typified by 2 years or more of post-secondary study |
| 2. Knowledge of engineering sciences: Breadth and depth of education; type of knowledge | Apply knowledge of mathematics, science, engineering fundamentals and an engineering specialisation to the conceptualization of engineering models | Apply knowledge of mathematics, science, engineering fundamentals and an engineering specialisation to defined and applied engineering procedures, processes, systems or methodologies | Apply knowledge of mathematics, science, engineering fundamentals and an engineering specialisation to wide practical procedures and practices |
| 3. Problem analysis: Complexity of analysis | Identify, formulate, research literature and solve complex engineering problems reaching substantiated conclusions using first principles of mathematics and engineering sciences | Identify, formulate, research literature and solve broadly defined engineering problems reaching substantiated conclusions using analytical tools appropriate to their discipline or area of specialisation | Identify and solve well-defined engineering problems reaching substantiated conclusions using codified methods of analysis specific to their field of activity |
| 4. Design/development of solutions: Breadth and uniqueness of engineering problems | Design solutions for complex engineering problems and design systems, components or processes that meet specified needs with appropriate consideration for public health and safety, cultural, societal and environmental considerations | Design solutions for broadly defined engineering technology problems and contribute to the design of systems, components or processes to meet specified needs with appropriate consideration for public health and safety, cultural, societal and environmental considerations | Design solutions for well-defined technical problems and assist with the design of systems, components or processes to meet specified needs with appropriate consideration for public health and safety, cultural, societal and environmental considerations |

| | | | |
|---|---|---|---|
| 5. Investigation: Breadth and depth of investigation and experimentation | Conduct investigations of complex problems including design of experiments, analysis and interpretation of data and synthesis of information to provide valid conclusions | Conduct investigations of broadly defined problems; locate, search and select relevant data from codes, data bases and literature, design and conduct experiments to provide valid conclusions | Conduct investigations of well-defined problems; locate and search relevant codes and catalogues, conduct standard tests and measurements |
| 6. Modern tool usage: Level of understanding of the appropriateness of the tool | Create, select and apply appropriate techniques, resources and modern engineering tools, including prediction and modelling, to complex engineering activities, with an understanding of the limitations | Select and apply appropriate techniques, resources and modern engineering tools, including prediction and modelling, to broadly defined engineering activities, with an understanding of the limitations | Apply appropriate techniques, resources and modern engineering tools to well-defined engineering activities, with an awareness of the limitations |
| 7. Individual and team work: Role in and diversity of team | Function effectively as an individual, and as a member or leader in diverse teams and in multi-disciplinary settings | Function effectively as an individual, and as a member or leader in diverse technical teams | Function effectively as an individual, and as a member in diverse technical teams |
| 8. Communication: Level of communication according to type of activities performed | Communicate effectively on complex engineering activities with the engineering community and with society at large, such as being able to comprehend and write effective reports and design documentation, make effective presentations and give and receive clear instructions | Communicate effectively on broadly defined engineering activities with the engineering community and with society at large, by being able to comprehend and write effective reports and design documentation, make effective presentations and give and receive clear instructions | Communicate effectively on well-defined engineering activities with the engineering community and with society at large, by being able to comprehend the work of others, document their own work and give and receive clear instructions |

(continued)

**Table 5** (continued)

| Characteristic | …for Washington Accord graduate | …for Sydney Accord graduate | …for Dublin Accord graduate |
|---|---|---|---|
| 9. The Engineer and Society: Level of knowledge and responsibility | Demonstrate understanding of the societal, health, safety, legal and cultural issues and the consequent responsibilities relevant to engineering practice | Demonstrate understanding of the societal, health, safety, legal and cultural issues and the consequent responsibilities relevant to engineering technology practice | Demonstrate knowledge of the societal, health, safety, legal and cultural issues and the consequent responsibilities relevant to engineering technician practice |
| 10. Ethics: No differentiation in this characteristic | Understand and commit to professional ethics and responsibilities and norms of engineering practice | Understand and commit to professional ethics and responsibilities and norms of engineering technology practice | Understand and commit to professional ethics and responsibilities and norms of technician practice |
| 11. Environment and Sustainability: No differentiation in this characteristic | Understand the impact of engineering solutions in a societal context and demonstrate knowledge of and need for sustainable development | Understand the impact of engineering solutions in a societal context and demonstrate knowledge of and need for sustainable development | Understand the impact of engineering solutions in a societal context and demonstrate knowledge of and need for sustainable development |
| 12. Project management and finance: Level of management required for differing activities | Demonstrate a knowledge and understanding of management and business practices, such as risk and change management, and understand their limitations | Demonstrate an awareness and understanding of management and business practices, such as risk and change management, and understand their limitations | Demonstrate an awareness of management and business practices, such as risk and change management |
| 13. Life-long learning: No differentiation in this characteristic | Recognise the need for, and have the ability to engage in independent and life-long learning | Recognise the need for, and have the ability to engage in independent and life-long learning | Recognise the need for, and have the ability to engage in independent and life-long learning |

## References


Director, S. W., Khosla, P. K., Rohrer, R. A., & Rutenbar, R. A. (1995). Reengineering the curriculum: Design and analysis of a new undergraduate electrical and computer engineering degree at Carnegie Mellon University. *Proceedings of the IEEE, 83*(9), 1246–1259.

Engineers Mobility Forum. (2007). *The international register of professional engineers, constitution.* Revised June 2007, http://www.ieagreements.org/

Institution of Engineers Australia. (1996). Changing the culture: Engineering education into the future. *Australia institution of engineers review report, 1996.*

International Engineering Alliance. (2005). *Graduate attributes and professional competencies*, Ver 1.1–13, June 2005, http://www.ieagreements.org/

International Engineering Alliance. (2007). *Rules and Procedures*, August 2007, http://www.ieagreements.org/

Lang, J. D., Cruse, X., McVey, F. D., & McMasters, J. (1999). Industry expectations of new engineers: A survey to assist curriculum designers. *Journal of Engineering Education, 88*(1), 43.

Miksad, R. et al. (1996). Breadth vs. depth. *ASEE Prism, March 1996*, 48.

van Valkenberg, M. E. (1990). An engineering curriculum for the future. *IEEE Communications Magazine, 28*(12), 12–15.

Wenk, E. (1997). Teaching engineering as a social science. *ASEE Prism, December 1997*, 24–28.

# Quality Assurance in the Preparation of Technical Professionals: The ABET Perspective

**George D. Peterson**

**Abstract** Accreditation provides public recognition that an educational institution or program has met certain standards or criteria regarding students, curriculum, faculty, facilities, and support, ensuring a quality educational experience. Quality Assurance organizations generally adhere to the fundamental principles that accreditation is a nongovernmental process of peer review that educational institutions or programs may voluntarily undergo to determine if they are in compliance with standards set by the organization. There are two types of accreditation in USA – institutional and programmatic. The accreditation process requires a self-study undertaken by the institution, department, or program on a continuous basis, with frequent reviews, normally site visits, taking place every 5 to 10 years. Accreditation and certification are sometimes confused. In general, institutions, departments, and programs are accredited, and individuals are certified.

## Introduction

In the United States, ABET, Inc.[1] (formerly the Accreditation Board for Engineering and Technology) is responsible for the programmatic accreditation of educational programs in applied science, computing, engineering, and engineering technology fields. These programs are not ranked; they either receive accreditation or are denied. Programmatic accreditation plays a vital role in the Quality Assurance of

G.D. Peterson (✉)
ABET, Inc., 111 Market Place, Suite 1050, Baltimore, MD 21202, USA
e-mail: gpeterson@abet.org

[1] ABET's role in the accreditation process began in 1932 when the Engineers' Council for Professional Development (ECPD) was formed to promote the status of the engineering profession and enhance the quality of engineering education. In 1980, the ECPD became the Accreditation Board for Engineering and Technology focusing its efforts on the accreditation of educational programs. ABET is a federation of professional engineering and technical societies, and representatives from these societies, who are practicing professionals from industry and academe, form the body of ABET through its Board of Directors and four working commissions.

A.S. Patil and P.J. Gray (eds.), *Engineering Education Quality Assurance: A Global Perspective*,
DOI 10.1007/978-1-4419-0555-0_5, 

professional education. It is the primary means by which colleges, universities, and programs assure quality to students and the public. It facilitates the smooth transfer of courses and programs among institutions. While also greatly important to the graduates' professional education and career aspirations, programmatic accreditation serves as the bridge linking industry, the professions, and the colleges and universities. Equally important are the global mutual recognition agreements (MRAs) among agencies in countries throughout the world designed to affirm the substantial equivalence of accreditation systems and, thereby, encourage the recognition of graduates' education from these systems as appropriate for entry into the profession.

## Accreditation in the United States

Accreditation is the public recognition that an educational institution or program has met certain standards or criteria regarding students, curriculum, faculty, facilities, and support. In the United States, accreditation is a nongovernmental, peer-review process. It is voluntary on the part of the institution; however, because institutional accreditation status is tied to eligibility for federal student aid in USA, very few institutions forgo accreditation. At the program level, accreditation in some disciplines, primarily those related to public health, safety, and welfare, such as social work, architecture, medicine, and law, may be mandated by state legislatures and/or the professions themselves. In many such fields, graduation from an accredited program is among the requirements for professional practice. Engineering is one of these fields. As a result, approximately 98% of US engineering programs are accredited.

General accreditation review cycles in USA range from 6 to 10 years, with some accreditation bodies requiring annual reports in the interim. General reviews require the institution or program to submit a detailed self-study report, which documents compliance with the accreditation criteria. Following review of the report, the accreditor sends a team of evaluators to the university campus to investigate further areas of compliance that are not easily documented on paper. These may include laboratories, libraries, and other facilities. The evaluators also interview students, faculty, and others and review student work and course materials. In addition, they may observe classes. The evaluators are usually educators or practitioners in the disciplines under review or educational administrators in the case of institutional reviews. Often, observers will accompany evaluation teams. In engineering reviews, these observers are primarily representatives of state boards of professional licensure and registration. With few exceptions, evaluators are unpaid volunteers.

The accreditation process in USA is very labor-intensive, requiring many hours of effort on the part of both the institutions and the evaluators. The time involved in self-study preparation is significant. For evaluators, there is training, self-study review, and campus travel ranging from 3 to 5 days. Evaluation team leaders carry on the process for months following the visit, compiling, editing, and ultimately presenting to peer team leaders or, within some accreditation agencies, a board of directors, the team's recommendations for an accreditation action. ABET estimates that its evaluators and team leaders dedicate 2–3 weeks per year to the accreditation process.

Current issues in US accreditation include moving to an outcomes-based evaluation paradigm, ensuring the accountability of accreditors and institutions, managing the accreditation-related workload of institutions and volunteers, and responding to the need to evaluate an increasing number of nontraditional institutions and programs, i.e., alternative delivery systems. The Council for Higher Education Accreditation (CHEA) is the recognition body for US accreditors and, so, the agency most involved in working these issues on the largest scale.

## History of ABET

Prados (2007) recounts the history of ABET, which was established in 1932 as the Engineers' Council for Professional Development (ECPD). ECPD was formed to fill the apparent need for a *joint program for upbuilding engineering as a profession*, a need determined through surveys conducted by professional engineering societies in the 1920s. The ECPD's original focus was in four areas:

- *Guidance*. Supplying information to engineering students and potential students
- *Training*. Developing plans for personal and professional development
- *Education*. Appraising engineering curricula and maintaining a list of accredited curricula
- *Recognition*. Developing methods whereby individuals could achieve recognition by the profession and the general public

Seven engineering societies founded the organization and contributed to its original direction and focus: The American Society of Civil Engineers (ASCE), the American Institute of Mining and Metallurgical Engineers (now the American Institute of Mining, Metallurgical, and Petroleum Engineers), the American Society of Mechanical Engineers (ASME), the American Institute of Electrical Engineers (now IEEE), the Society for the Promotion of Engineering Education (now the American Society for Engineering Education), the American Institute of Chemical Engineers (AIChE), and the National Council of State Boards of Engineering Examiners (now NCEES).

Within its first year of existence, ECPD had begun developing itself as an accreditation agency; in 1936, ECPD evaluated its first engineering degree programs. Ten years later, the council began evaluating engineering technology degree programs. By its 15th year, ECPD had accredited 580 undergraduate engineering curricula at 133 institutions.

In 1980, ECPD was renamed the Accreditation Board for Engineering and Technology (ABET) to more accurately describe its emphasis on accreditation. Three years later, ABET created the Related Accreditation Commission, now known as the Applied Science Accreditation Commission (ASAC). In 1985, in response to the anticipated boom in computer science education, ABET helped establish the Computing Sciences Accreditation Board (now CSAB). CSAB is now one of the ABET's largest member societies with more than 300 accredited programs.

In 2005, ABET formally changed its name from the *Accreditation Board for Engineering and Technology* to ABET, Inc. This allows the organization to continue its activities under the name that represents leadership and quality in accreditation for the public while reflecting its broadening into additional areas of technical education.

Currently, ABET accredits approximately 2,800 applied science, computing, engineering, and technology programs at more than 650 colleges and universities. Each year, over 1,500 volunteers from its now 30 member societies actively contribute to ABET's goals of leadership and Quality Assurance in technical education, serving as program evaluators, committee and council members, commissioners, and board representatives.

ABET has been recognized by the CHEA since 1997.

## Structure of ABET

ABET is owned and operated by 30 professional and technical societies. Those societies appoint members to the board of directors, nominate members of the accreditation commissions, and select and assign program evaluators. All of these are volunteer positions. The societies also establish certain program accreditation criteria for their disciplines.

The role of ABET's Board of Directors is to set policy and strategy for the organization. The role of the accreditation commissions is to develop general accreditation criteria, set and carry out accreditation procedures, and accredit programs. The work of these bodies is supported by a paid executive director and a small professional headquarters staff.

ABET volunteers are mid- to late-career professionals in the disciplines that ABET accredits. There is also a small number of public representatives (individuals not working in or related to the ABET disciplines) on its commissions and board. ABET strives to maintain equal representation among academicians and practicing professionals from industry, government, and private practice. Typically, volunteers are split 60/40 among the academic and nonacademic arenas, respectively.

## ABET's Accreditation Criteria

In 1997, following nearly a decade of development, ABET adopted Engineering Criteria 2000 (*EC2000*), considered at the time a revolutionary approach to accreditation criteria (Accreditation Board for Engineering and Technology, Inc. 1997). The revolution of *EC2000* was its focus on what is learned rather than what is taught. At its core was the call for a continuous quality improvement process informed by the specific mission and goals of individual institutions and programs. Lacking the inflexibility of earlier accreditation criteria, *EC2000* allowed ABET to enable program innovation, as well as encourage new assessment processes and subsequent program improvement. Listed below is an excerpt from the 2008 to 2009 *Criteria for Accrediting Engineering Programs* (Accreditation Board for Engineering and Technology, Inc. 2007).

## *General Criteria for Baccalaureate-Level Programs*

### Criterion 1: Students

The program must evaluate student performance, advise students regarding curricular and career matters, and monitor student's progress to foster their success in achieving program outcomes, thereby enabling them as graduates to attain program objectives.

The program must have and enforce policies for the acceptance of transfer students and for the validation of courses taken for credit elsewhere. The program must also have and enforce procedures to assure that all students meet all program requirements.

### Criterion 2: Program Educational Objectives

Each program for which an institution seeks accreditation or reaccreditation must have in place:

(a) Published educational objectives that are consistent with the mission of the institution and these criteria
(b) A process that periodically documents and demonstrates that the objectives are based on the needs of the program's various constituencies
(c) An assessment and evaluation process that periodically documents and demonstrates the degree to which these objectives are attained

### Criterion 3: Program Outcomes

Engineering programs must demonstrate that their students attain the following outcomes:

(a) An ability to apply knowledge of mathematics, science, and engineering
(b) An ability to design and conduct experiments, as well as to analyze and interpret data
(c) An ability to design a system, component, or process to meet desired needs within realistic constraints such as economic, environmental, social, political, ethical, health and safety, manufacturability, and sustainability

(d) An ability to function on multidisciplinary teams
(e) An ability to identify, formulate, and solve engineering problems
(f) An understanding of professional and ethical responsibility
(g) An ability to communicate effectively
(h) The broad education necessary to understand the impact of engineering solutions in a global, economic, environmental, and societal context
(i) A recognition of the need for, and an ability to engage in lifelong learning
(j) A knowledge of contemporary issues
(k) An ability to use the techniques, skills, and modern engineering tools necessary for engineering practice.

Program outcomes are outcomes (a)–(k) plus any additional outcomes that may be articulated by the program. Program outcomes must foster attainment of program educational objectives.

There must be an assessment and evaluation process that periodically documents and demonstrates the degree to which the program outcomes are attained.

## Criterion 4: Continuous Improvement

Each program must show evidence of actions to improve the program. These actions should be based on available information, such as results from Criteria 2 and 3 processes.

## Criterion 5: Curriculum

The curriculum requirements specify subject areas appropriate to engineering but do not prescribe specific courses. The faculty must ensure that the program curriculum devotes adequate attention and time to each component, consistent with the outcomes and objectives of the program and institution. The professional component must include:

(a) One year of a combination of college level mathematics and basic sciences (some with experimental experience) appropriate to the discipline
(b) One and one-half years of engineering topics, consisting of engineering sciences and engineering design appropriate to the student's field of study. The engineering sciences have their roots in mathematics and basic sciences but carry knowledge further toward creative application. These studies provide a bridge between mathematics and basic sciences on the one hand and engineering practice on the other. Engineering design is the process of devising a system, component, or process to meet desired needs. It is a decision-making process (often iterative), in which the basic sciences, mathematics, and the engineering sciences are applied to convert resources optimally to meet these stated needs.
(c) A general education component that complements the technical content of the curriculum and is consistent with the program and institution objectives.

Students must be prepared for engineering practice through a curriculum culminating in a major design experience based on the knowledge and skills acquired in earlier course work and incorporating appropriate engineering standards and multiple realistic constraints.

**Criterion 6: Faculty**

The faculty must be of sufficient number and must have the competencies to cover all of the curricular areas of the program. There must be sufficient faculty to accommodate adequate levels of student–faculty interaction, student advising and counseling, university service activities, professional development, and interactions with industrial and professional practitioners, as well as employers of students.

The program faculty must have appropriate qualifications and must have to demonstrate sufficient authority to ensure the proper guidance of the program and to develop and implement processes for the evaluation, assessment, and continuing improvement of the program, its educational objectives, and outcomes. The overall competence of the faculty may be judged by factors such as education, diversity of backgrounds, engineering experience, teaching effectiveness and experience, ability to communicate, enthusiasm for developing more effective programs, level of scholarship, participation in professional societies, and licensure as Professional Engineers.

**Criterion 7: Facilities**

Classrooms, laboratories, and associated equipment must be adequate to safely accomplish the program objectives and provide an atmosphere conducive to learning. Appropriate facilities must be available to foster faculty–student interaction and to create a climate that encourages professional development and professional activities. Programs must provide opportunities for students to learn the use of modern engineering tools. Computing and information infrastructures must be in place to support the scholarly activities of the students and faculty and the educational objectives of the program and institution.

**Criterion 8: Support**

Institutional support, financial resources, and constructive leadership must be adequate to assure the quality and continuity of the program. Resources must be sufficient to attract, retain, and provide for the continued professional development of a well-qualified faculty. Resources also must be sufficient to acquire, maintain, and operate facilities and equipment appropriate for the program. In addition, support personnel and institutional services must be adequate to meet program needs.

### Criterion 9: Program Criteria

Each program must satisfy applicable Program Criteria (if any). Program Criteria provide the specificity needed for interpretation of the baccalaureate level criteria as applicable to a given discipline. Requirements stipulated in the Program Criteria are limited to the areas of curricular topics and faculty qualifications. If a program, by virtue of its title, becomes subject to two or more sets of Program Criteria, then that program must satisfy each set of Program Criteria; however, overlapping requirements need to be satisfied only once.

## Evaluating the Impact of Outcomes-Based Accreditation

In 2002, ABET commissioned the Center for the Study of Higher Education at Pennsylvania State University to undertake a three-and-a-half-year study to assess whether the implementation of the new outcomes-based evaluation criteria was having the intended effects. *Engineering Change: A Study of the Impact of EC2000* (Lattuca et al., 2006) was designed to answer two primary questions:

- What impact, if any, has *EC2000* had on student-learning outcomes in ABET-accredited programs and institutions?
- What impact, if any, has *EC2000* had on organizational and educational policies and practices that may have led to improved student-learning outcomes?

The study was based on data collected by Penn State from 40 institutions, 147 programs, 1,243 faculty, 5,494 graduates of 1994, 4,330 graduates of 2004, and 1,622 employers. As highlighted in the *Engineering Change: A Study of the Impact of EC2000 – Executive Summary,* key findings of this study include the following:

- Greater emphasis on professional skills and active learning after EC2000
- High levels of faculty support for continuous improvement
- 2004 graduates are better prepared than their 1994 counterparts
- Professional skills are gained, while technical skills are maintained
- Changes in programs and student experiences empirically linked to higher performance

The strengths of an outcomes-based accreditation model are now broadly recognized, as evident from increasing adoption by institutional and specialized accreditors worldwide. In addition, the program outcomes of ABET criteria (a)–(k) are being replicated by engineering accreditors in multiple jurisdictions.

## International Quality Assurance Activities

ABET's international activities are a continuously growing area of the organization. Currently, there are three major areas of international activity: establishment of MRAs, consultancy services for accreditation and higher education agencies, and accreditation of nondomestic programs.

## *Mutual Recognition Agreements*

MRAs are designed to affirm the substantial equivalence of accreditation systems and, thereby, encourage the recognition of graduates' education from these systems as appropriate for entry into the profession (Phillips et al. 2000). The first MRA, between ABET and the Canadian Council of Professional Engineers (CCPE), now known as Engineers Canada, was signed in 1980. Although the criteria used for CCPE accreditation differ from ABET's, the outcome – graduates who are prepared to enter the profession – is the same. As a result of the success of the MRA between ABET and CCPE, interest was expressed by other organizations in establishing similar agreements. This led to the formation of the Washington Accord in 1989. Established originally as the Six Nation Agreement, the Washington Accord has formed the basis of what is now called the International Engineering Alliance, consisting of six international agreements (the Washington, Sydney, and Dublin Accords and the Engineers Mobility Forum, the Engineering Technologist Mobility Forum, and the Asia Pacific Economic Cooperation) covering the mutual recognition of engineering and technology education and extending to the professional competence of the practicing engineer, technologist, and technician (Peterson and Hanrahan 2006).

The founding signatories of the Washington Accord are from the jurisdictions of the United States, the United Kingdom, Canada, Australia, New Zealand, and Ireland. The current Washington Accord consists of 11 signatories.

ABET's presence in the Washington Accords extends beyond its role as a founding signatory. From 2001 to 2007, ABET served as the accord's Secretariat, responsible for the management and administration of the Rules and Procedure. Currently, the Chair of the Washington Accord is a representative of ABET.

## *Consultancy Services*

ABET often receives requests from Quality Assurance organizations or higher education authorities outside USA seeking assistance in developing or enhancing their accreditation systems. This assistance is often initiated by the signing of a Memorandum of Understanding (MOU). By signing an MOU, the parties agree to *collaborate on matters related to technical education and Quality Assurance activities.* ABET has negotiated MOUs with Quality Assurance agencies in approximately 16 countries and regions throughout the world. Typical activities following an MOU include facilitating the exchange of representatives to observe signatories' accreditation evaluation activities and sharing best practices; promoting accreditation principles and facilitating the training of evaluators; exchanging information, including documents, procedures, and surveys concerning accreditation processes and higher education Quality Assurance; and exploring the feasibility of mutual recognition based on the monitoring and assessment of their respective accreditation systems.

The Western Hemisphere Initiative, an MOU established in 2002, is designed to promote cooperation in the enhancement of Quality Qssurance systems within North America, Central America, the Caribbean, and South America. Members of the Western Hemisphere Initiative include ABET, the Council for Higher Education Accreditation (CHEA), the Consejo de Acreditación de la Ensenanza de la Ingeniería (CACEI) of Mexico, Engineers Canada, and, most recently, the Instituto Para la Calidad en la Acreditación en las Carreras de Ingenería y Tecnología (ICACIT) of Peru.

## *Nondomestic Accreditation*

In the recent past, ABET evaluated nondomestic engineering programs, upon request from institutions, to determine if they were substantially equivalent to accredited programs in the United States. *Substantial equivalency recognition* was available only to programs in jurisdictions where accreditation bodies were not signatories of the Washington Accord. The evaluation used the same criteria for accreditation in the United States, while making some modifications for educational and cultural differences. Over the past few years, ABET has conducted substantial equivalency visits in 17 countries and granted recognition to approximately 160 programs at 31 institutions.

In October 2006, the ABET Board of Directors granted approval to accredit programs outside the United States and to phase out substantial equivalency evaluations. The extensive international experience with substantial equivalency visits, among other activities, helped ABET to prepare with the necessary tools to transition successfully into nondomestic accreditation. The first international accreditation visits were conducted in fall 2007. Twenty-one programs in five countries were evaluated. Their accreditation actions were determined in the summer of 2008.

ABET's venture into nondomestic accreditation is important for several reasons. In today's global economy, there is increasing demand for cross-border quality recognition. ABET accreditation provides a mechanism for such recognition by enforcing accepted technical educational standards. Programs that comply with ABET's policies and criteria can be expected to produce graduates qualified to enter the applied science, computing, engineering, and technology professions in any country on earth. This is critical, as the safety of the public is dependent on qualified practitioners in these areas. Thus, accreditation facilitates the mobility of technical professionals. As cultures around the world continue to diversify and expand their global reach, it is important that licensing boards and employers are confident in the quality of education that graduates receive. Individual practitioners accrue the benefit of global recognition of their credentials.

## Conclusions

ABET is currently engaged in several strategy-making pursuits. The first is making accreditation available to all requestors. As educational delivery mechanisms change and demand for accreditation crosses more disciplines and borders, so much accreditation evaluation policies and processes change. ABET is currently working to respond to an increasing volume of requests from nontraditional and nondomestic institutions and programs and is moving toward changes in its processes as a result. Another is ensuring that all relevant constituencies are represented within ABET, Inc. It is clear that some entities some that benefit from and participate in ABET accreditation are absent from the governance function of the organization. ABET is currently exploring new governance structures that would address this challenge.

Quality Assurance of technical education will be challenged in the coming years by the changing dynamics of education: globalization, alternative delivery systems (online education), emerging technologies, and blurring of the boundaries between traditional disciplines. Overlying these challenges is the evolution of the traditional 4-year degree learning structure to a mosaic of knowledge, skills, and competencies acquired through a diverse spectrum of sources, for example, self-learning, online courses, internships, and apprenticeships, as well as the residential bricks and mortar campus. The measure of competence may shift from what the graduate was taught to what the graduate knows and, as a result, the emphasis on Quality Assurance may shift from accreditation of programs to credentialing the individual.

## References

Accreditation Board for Engineering and Technology, Inc. (1997). *A vision for change: A summary report of the ABET/NSF industry workshops*. Baltimore, MD: ABET, Inc.

Accreditation Board for Engineering and Technology, Inc. (2007). *Criteria for accrediting engineering programs: Effective for evaluations during the 2008–2009 accreditation cycle.* Baltimore MD: ABET, Inc. Accessed January 2009. http://www.abet.org/Linked%20 Documents-UPDATE/Criteria%20and%20PP/E001%2008-09%20EAC%20Criteria%20 12-04-07.pdf

Lattuca, L. R., Terenzini, P. T., Volkwein, J. F., et al. (2006). *Engineering change: A study of the impact of EC2000*. Baltimore, MD: ABET, Inc.

Peterson, G. D., & Hanrahan, H. (2006). *Graduate attributes and professional competency profiles for the engineer, engineering technologist, and engineering technical.* Development of curricula for the global engineer, 2006 ASEE Global Colloquium on Engineering Education, Rio de Janeiro, October 10, 2006.

Phillips, W. M., Peterson, G. D., & Aberle, K. B. (2000). *Quality Assurance for Engineering Education in a Changing World.* International Journal for Engineering Education, *16*(3), 97–103. Accessed January 2008. http://www.ijee.dit.ie/articles/Vol16-2/Ijee1154.pdf

Prados, J. W. (ed). (2007). *A proud legacy of quality assurance in the preparation of technical professionals: ABET 75th anniversary retrospective*. Baltimore, MD: ABET, Inc.

# Issues and Challenges (Country/Institutional Perspectives)

# Quality Assurance in Engineering Education and Modernization of Higher Education in Russia

A. Chuchalin, O. Boev, and A. Kriushova

**Abstract** The paper discusses the Russian Federation system of higher education regarding the implementation of the key aspects of the Bologna process. Some recent legislative initiatives aimed at the modernization of higher education in Russia in accordance with the main Bologna principles are also presented. The issues of Quality Assurance and, in particular, differences between state and professional accreditation are described. Most importantly, the role of the Russian Association for Engineering Education in developing the national system for professional accreditation in engineering education and its contribution to establishing a pan-European system for Quality Assurance in engineering education being created within the European Higher Education Area (EHEA) are described.

## Introduction

Though recognized for its rich traditions, Russian higher education has long remained in isolation from the world community as a result of its being under the restraints of a closed planned economy. Through a system of *state orders* (*goszakaz*) the government introduced a range of programmes for training specialists and provided for uniform curricula in higher education institutions (HEIs). The transition to an internal market economy and the integration of the Russian Federation into the world economy has necessitated the revision of its approaches to higher education

As a result, Russian higher education initiated a range of modernizing processes. These included the introduction of a multilevel educational system, competition with other Russian and foreign educational establishments, increased independence of students, and the development of new educational technologies. In addition,

O. Boev (✉),
Accreditation Center, Russian Association for Engineering Education, Office 328,
30 Lenin Avenue, Tomsk 634050, Russia
e-mail: ovb@ac-raee.ru

A.S. Patil and P.J. Gray (eds.), *Engineering Education Quality Assurance: A Global Perspective*,
DOI 10.1007/978-1-4419-0555-0_6, © Springer Science+Business Media, LLC 2009

Russian universities' academic programmes have had to become flexible to respond to market demands and needs.

Integration into the world education community while preserving the traditions and achievements of the Russian system of higher education is one of the fundamental principles in the state policy in education (Government of the Russian Federation 2005). In September 2003, the Russian Federation officially joined the Bologna Process and became a full participant in creating the European Higher Education Area (EHEA). Along with the other members of the EHEA, the Russian Federation made the commitment to transform the *national system of higher education* in accordance with the main principles of the Bologna Process by the year 2010.

Joining the Bologna Process and the integration of the Russian Federation into EHEA are key priorities in the modernization and development of the *national higher education system.* To guide the integration of the Russian Federation into EHEA, the Ministry of Education and Science has created a special working group along with a plan for implementing the Bologna principles in Russian Federation higher education (Ministry of Education and Science 2004, 2005).

## Two-Tier System

In 1992, the Committee for Higher Education of the Ministry of Science issued a legal decree on the *Introduction of the Multi-level Structure for Higher Education in the Russian Federation* (Committee for Higher Education 1992). In accordance with the decree, the 4-year Bachelors degree programmes were introduced as *First Cycle Degree* (FCD) programmes. Upon completion of FCD programmes, graduates have a choice either to enter the labor market or to continue their studies to obtain the *Second Cycle Degree* (SCD). The SCD was awarded to the graduates who had completed the 5-year Diploma Specialist (one more year of studies after obtaining the FCD) or 2-year Masters programmes.

New versions of the Federal Laws on *Education* and on *Higher Professional Education* adopted by the State Duma of the Russian Federation in December 2007 fixed the two-tier system of higher education, with 4-year Bachelors programmes and successive 2-year Masters programmes. The new laws canceled the transition from Bachelor to Diploma Specialist programmes. It is worth noting that integrated programmes leading directly to the SCD remained in certain areas, including some engineering disciplines.

## State Educational Standards

In mid 1990s, the Ministry of Education introduced the *State Educational Standards for Higher Professional Education* (SES), first and second versions, dated 1994 and 2000, respectively (Ministry of Education 1994, 2000). These standards set minimum requirements for programme content and quality as well as the time allocated for accomplishing the programme and basic specialists' qualifications. Though the education system in Russia remains centralized through governmental control over the structure and content of programmes, the HEIs were granted some academic freedom in programme design.

Still the Standards define from 60 to 70% of programme content and have federal, national, and regional HEI components. Therefore, there were several weaknesses in these Standards that needed to be eliminated including the lack of freedom in programme design on the part of institutions and in the choice of electives by students, the strict order of courses in the curricula and the linear organization of the learning process, insufficient time allocated for students' independent work and self-study, and weak control over the assessment of students' achievement of learning outcomes.

Both internal and external factors necessitated the revision of the former Standards. Over the past few years, work on the elaboration of a new (third) version of the Standards was been initiated by a group including a wide range of academics and representatives of profession. The framework for the new *Federal Educational Standards* (FES) was approved by the Russian Ministry of Education and Science in February 2007.

The principal difference of the new version of the Standards from the previous ones is the *outcomes-based approach.* It is assumed that the new Standards will define the general requirements for graduates' competencies, both professional and personal (transferable). The definition of *learning outcomes* (knowledge, skills, and attitudes) that students should demonstrate upon graduation and the assessment of their fulfilment is of essential importance in the third version of the Standards. The new approach assumes active involvement of the professional community and employers in the formulation of general as well as specific *learning outcomes.*

The use of the *outcomes-based approach* grants HEIs more academic freedom in the design of their programmes and curricula. Although programmes must assure that their graduates gain the required set of competencies, the ways of achieving these competencies may be different. Undoubtedly, the *outcomes-based approach* of the new FES will facilitate both dialogue between academic and professional communities and the recognition of the qualifications and degrees.

The other legislative initiative of the FES directs the differentiation of two tracks within the System of Higher Education. The first is the new practically oriented Bachelors and Masters Degree programmes. These are to be established in parallel with the existing research-oriented programme. The establishment of the practically oriented programmes will facilitate the specification of engineering education.

## Quality Assurance and Accreditation

Issues related to higher education Quality Assurance became crucial in mid-1990s, when HEIs were given more flexibility in programme design and the number of HEIs and programmes began to increase. To ensure the quality of higher education, a procedure of accreditation was implemented through the 1992 *Federal Law on Education* (Government of the Russian Federation 1992). According to the Law, accreditation exists at two levels, *state* (run by the Ministry of Education of the Russian Federation) and *professional* (run by public professional organizations). State accreditation is *institutional* while professional accreditation deals with educational *programmes*.

The state accreditation system is an integrated assessment of HEIs based on a comprehensive analysis of HEI activities. It includes three procedures for *licensing*, *attestation*, and state *accreditation* (see Fig. 1).

*Licensing* certifies that HEIs' facilities, financial support, and resources, including information, are adequate to meet *state requirements*. The purpose of *licensing* is to establish the right of a HEI to provide educational services. *Attestation* is the establishment of *equivalency* between the content, level, and quality of the education offered and the requirements set by the *SES*. *State accreditation* grants to the HEI the right to award state degrees and confirms the status of the HEI (academy, institute, or university). The *Certificate of State Accreditation* is issued for a 5-year period.

Although the *state accreditation* is *institutional*, that is, it evaluates HEIs in general, *professional accreditation* focuses on the assessment of the content and quality of particular educational programmes. *Professional accreditation* relies on accreditation criteria that are more rigorous than the requirements of the *State*

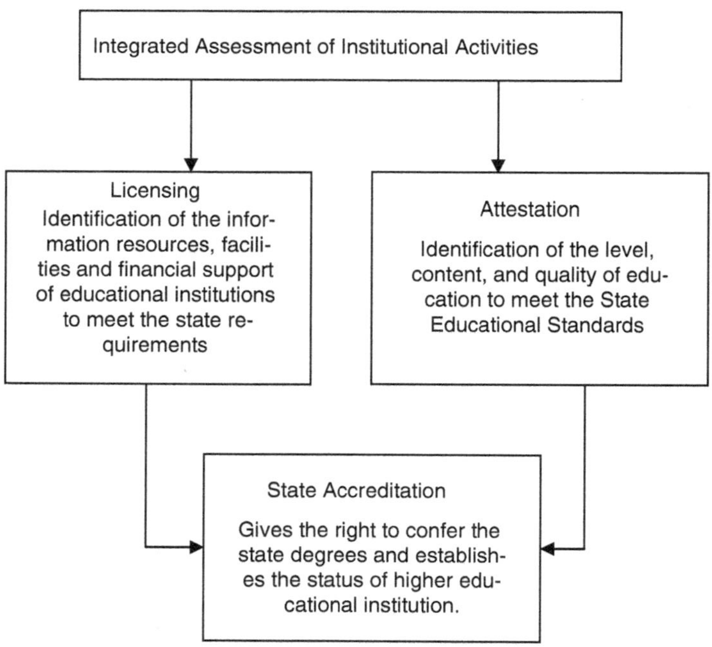


**Fig. 1** Russian state accreditation system

*Educational Standard*. In accordance with the *Federal Law on Education*, *professional accreditation* is the responsibility of public professional organizations. The Russian Association for Engineering Education (RAEE, http://www.aeer.ru) is responsible for professional accreditation in engineering and technology. The system for *professional accreditation* is well developed in engineering education as described next.

## National System of Quality Assurance in Engineering Education

*Quality Assurance* of higher education, that is, the development of effective systems for *Quality Assurance* within HEIs, nationally, and Europe-wide is a cornerstone of the EHEA. In accordance with the *Berlin Communiqué* (Bologna Process 2005), national systems for *Quality Assurance* must include external review of educational programmes as part of *accreditation* and *certification* procedures. Currently, engineering education has the only system for *professional accreditation* in the Russian Federation.

The RAEE was involved in developing the national system for professional accreditation in engineering education beginning in 1992 (Pokholkov et al. 2004). Up until 2001 the accreditation of the programmes in engineering and technology was carried out by the Independent Accreditation Center (IAC) founded by the RAEE. The IAC was the first nongovernmental body evaluating programmes of higher education in Russia. By 1999, the IAC had accredited a total of 34 Russian engineering and technology programmes. In mid-1990s, the Ministry of Education of the Russian Federation introduced the *state accreditation* of HEIs that overlapped considerably with IAC accreditation requirements. It was at that time the RAEE realized the need to adopt an *outcomes-based approach* instead of the input-oriented evaluation run by the IAC.

RAEE activities have been approved and supported both by the Ministry of Education and Science (Ministry of Education 2003) and by professional organizations, such as the 2004 agreement on cooperation between the RAEE, the Chamber of Commerce and Industry of the Russian, and the Union of Employers and Businessmen. At the national scale the RAEE principal aim is to enhance the quality of programmes in engineering and technology through their external evaluation. On a global scale, the RAEE strives for international recognition of accredited programmes and graduates' qualifications.

RAEE initiated the revision of the Russian engineering education *Quality Assurance* system based on two factors. The first is the importance to the engineering profession of professional community involvement in engineering programme evaluation. The second factor is international engineering agreements that are intended to make national *Quality Assurance* consistent with that of the world's leading engineering organizations. Revisions resulted in the elaboration of a new set of the working documents (the *outcomes-based* criteria and accreditation procedure, self-study manuals, expert guidelines) that are compatible with those of the

Washington Accord signatories – the world's leading organizations in the accreditation of programmes in engineering and technology.

The RAEE initiated revisions were encouraged by the leading Russian universities. These universities actively participated in the elaboration of the new approach to Quality Assurance and were involved in the establishment of the Accreditation Centre of the RAEE (http://www.ac-raee.ru) in November 2002. In April 2003, six leading technical universities, which contributed to the elaboration of the new approach, approved the accreditation criteria and procedures. To make the new accreditation system compatible internationally, the RAEE invited external observers to participate in pilot on-site visits and compare the accreditation process with that adopted by the Washington Accord signatories.

Over the past few years, the RAEE accreditation criteria and procedures have been essentially modified and improved based on valuable comments made by international observers. Since 2003, the RAEE has organized annual workshops for evaluators intended to prepare *Quality Assurance* experts to conduct accreditation reviews of engineering programmes using the new criteria and procedure. As the result of such workshops more than 100 experts have been certified. From 2003 to 2007 more than 50 engineering programmes in 20 Russian technical universities were accredited by the RAEE Accreditation Centre. About 40 programmes are currently preparing for accreditation. The majority of accredited programmes are 5-year integrated programmes leading to Specialist Diploma in engineering.

## International Aspects of the RAEE Accreditation Activity

The integration of the Russian Federation into the EHEA undoubtedly influenced the development of the national system for professional accreditation. In particular, in accordance with the main Bologna principles the RAEE revised its criteria so as to clearly differentiate between the FCD and the SCD programmes.

Since September 2004, the RAEE has actively participated in the elaboration of the common European system for accrediting engineering programmes set within the EUR-ACE (*EUR*opean-*AC*credited *E*ngineer) framework supported by the European Commission (see Chap. 3 by Augusti for a thorough discussion of EUR-ACE). The effort was carried out by the 14 partners including a wide range of organizations dealing with recognition and accreditation of engineering education. This *EUR-ACE Consortium* was made up of six European associations/networks and eight national agencies active in accreditation of engineering programmes.[1]

[1] Engineering Council UK (ECUK), Engineers Ireland, Commission des Titres d'Ingénieur (CTI); German Accreditation Agency Specialised in Accrediting Degree Programmes in Engineering, Informatics, the Natural Sciences and Mathematics (ASIIN); Portuguese Institution of Engineers (OE), Conference of Italian Engineering Deans (CoPI), Union of Associations of Civil Engineers of Romania (UAICR), and Russian Association for Engineering Education (RAEE).

The main objective of the EUR-ACE supported project was the elaboration of a framework for setting up a European system for accreditation of engineering education programmes at the First and Second Cycle levels. The work was divided into seven stages (including a preliminary stage and the extension of the project). The RAEE contributed to the Project at different stages: first, by presenting to the project working group the criteria and procedure for operating a national accreditation system in engineering education as well as information on systems of higher engineering education in Russia; second, by organizing three dissemination seminars in Russia; third, by providing recommendations, comments, and suggestions for refinement of the tentative *EUR-ACE Standards and Procedures*; and fourth, by participating along with Ireland and Turkey in the pilot test of the revised *EUR-ACE Standards and Procedures* for peer review of the engineering programmes.

As a result of the activities of the EUR-ACE project the *Consortium* members elaborated the *EUR-ACE Framework standards for the accreditation of engineering programmes* and agreed to establish an organization for implementation of the project proposals. In February 2006, the European Network for Accreditation of Engineering Education (ENAEE) was officially founded by 14 European engineering organizations including the RAEE.

In accordance with its mission the ENAEE is engaged in launching the decentralized European accreditation system of engineering study programmes. This system assumes that authorized national agencies accredit their programmes of study in accordance with criteria and procedure for accreditation that satisfy the EUR-ACE Framework Standards. This will enable them to apply the EUR-ACE label to accredited First or Second Cycle programmes.

In Russia, the *EUR-ACE Standards* were tested through the PRO-EAST project (*Promotion and Implementation of EUR-ACE Standards*). The project was carried out by the RAEE in cooperation with European partners such as UNIFI, SEFI, FEANI,[2] CoPI, ASIIN, and ECUK. The main outputs of the PRO-EAST project were as follows:

- RAEE working materials including the criteria and procedure for accreditation, guidelines for self-evaluation for HEIs, and manuals for programme evaluation experts that were reviewed and harmonized with the EUR-ACE *Standards* and ENQA *Standards* and *Guidelines* on Quality Assurance.
- Some of the RAEE experts were trained as members of international evaluation teams to conduct accreditation related to awarding the *EUR-ACE label*.
- Several Russian HEIs participated in the pilot test of the accreditation process conducted by international observers to award the *EUR-ACE label*.

In general, the PRO-EAST project helped disseminate the EUR-ACE *Standards* to the broader Russian engineering community. In addition, about 30 engineering programmes accredited by the RAEE Accreditation Centre have now been awarded the *EUR-ACE label*.

[2] University of Florence (UNIFI), Société Européenne pour la Formation d'Ingénieurs (SEFI), Fédération Européenne d'Associations Nationales d'Ingénieurs (FEANI).

In 2008, the RAEE accreditation activities (on-site visits for evaluation of FCD and SCD programmes as well as the meeting of the Accreditation Board) were reviewed by the international evaluation team. Based on the findings of the experts, the ENAEE Administrative Council reauthorized RAEE to award the *EUR-ACE label* for the next 5 years (until December 2013) to accredited FCD and SCD programmes. It is worth emphasizing that SCD programmes include both Master and Diploma Specialist Engineering programmes as they have identical requirements for programme outcomes.

The system of *EUR-ACE Labels*, accepted in EHEA, will facilitate the recognition of engineering graduates' qualifications throughout Europe. The RAEE participation in the EUR-ACE project and its follow-ups is an important step toward harmonizing Russian higher education and its national system of accreditation in engineering education with the common European system of *Quality Assurance* being created within the context of the Bologna process.

## Conclusions

Undoubtedly, the Russian system of higher education has been dramatically transformed over the past 15 years. Both internal and external factors forced the modernization of the national system of higher education. The reformation process that is currently underway should improve the quality of higher education as well as enhance the attractiveness and competitiveness of the national education system in international market of educational services.

The transition to a multilevel system should allow the Russian system of higher education to adapt to a market economy, increase the effective use of the financial resources allocated for education, and make Russian academic structures and degrees internationally compatible. In addition, the creation of an effective system of *Quality Assurance* and *accreditation* makes Russian universities more attractive by offering students the opportunity to complete internationally recognized degrees.

Understanding of problems existing in Russian higher education by all the constituencies (state, academic, and professional communities) and joint efforts in developing and modernizing the national system of higher education will facilitate our country the fulfilment of its obligations within EHEA and integration into the world education community.

## References

Bologna Process. (2005). Realising the European higher education area. Communiqué of the Conference of Ministers responsible for Higher Education in Berlin, 19 September 2003, from http://www.bologna-bergen2005.no/Docs/00-Main_doc/030919Berlin_Communique.PDF.

Committee for Higher Education. (1992). The Introduction of the multi-level structure for higher education in the Russian Federation. Decree of the Committee for Higher Education of the Ministry of Science on 13 March 1992, No. 13 (in Russian at: http://www.edu.ru).

Government of the Russian Federation. (1992). Federal Law On Education of 10 July 1992, No. 3266-1 (in Russian at: http://www.edu.ru).

Government of the Russian Federation. (2005). Federal programme on Development of Education for the years 2006–2010. Approved by Decree of the Government of the Russian Federation on 23 December 2005, No. 803.

Ministry of Education. (1994). State Educational Standards for Higher Professional Education, first version of 1994 (in Russian at: http://www.edu.ru).

Ministry of Education. (2000). State Educational Standards for Higher Professional Education, second version of 2000 (in Russian at: http://www.edu.ru).

Ministry of Education. (2003). On the development of a system for professional accreditation of programmes in engineering and technology. Order of the Ministry of Education on 27 May 2003, No. 2270.

Ministry of Education and Science. (2004). On a working group for the implementation of the Bologna principles in higher education. Decree of the Ministry of Education and Science on 25 October 2004, No. 100 (in Russian at: http://www.edu.ru).

Ministry of Education and Science. (2005). Plan on the implementation of the Bologna principles in higher professional education for 2005–2010. Approved by the Ministry of Education and Science on 15 February 2005, No. 40 (in Russian at: http://www.edu.ru).

Pokholkov, Y. P., Chuchalin, A. I., Boev, O. V., & Morozova, E. A. (2004). The accreditation of engineering programmes in Russia. *European Journal of Engineering Education, 29*(1), 163–169.

# Quality Assurance in Vietnam's Engineering Education

**Hao V. Le and Kim D. Nguyen**

**Abstract** The chapter first provides an overview of engineering education in Vietnam, and then describes the historical background of Quality Assurance in higher education in Vietnam and the role of government and its policies in regulating Quality Assurances practices at universities and colleges. Then, the current Quality Assurance scheme in higher education in Vietnam including engineering is presented. Strengths and weaknesses in implementing this scheme into engineering education are discussed. Initial efforts in applying international standards in engineering education, especially ABET criteria, at some universities are also considered. At last, recommendations for the future of Quality Assurance in engineering education in Vietnam are provided.

## Introduction

Engineering higher education in Vietnam has mainly developed since the 1950s with the establishment of Hanoi University of Technology in 1956, University of Technology – Ho Chi Minh City National University in 1957, and Thai Nguyen University of Mechanics and Electricity (formerly Thai Nguyen University of Technology) in 1965. Until late 2007 Vietnam has 172 universities including 22 military institutions. The authors have identified the following 18 major Vietnamese engineering and technology universities.

H.V. Le (✉)
Department of Educational Quality Assurance & Testing, Nha Trang University, 2 Nguyen Dinh Chieu, Nha Trang, Vietnam
e-mail: haolevan@yahoo.com

A.S. Patil and P.J. Gray (eds.), *Engineering Education Quality Assurance: A Global Perspective*,
DOI 10.1007/978-1-4419-0555-0_7, 

| Institution | Website |
|---|---|
| Hanoi University of Technology | http://www.hut.edu.vn |
| Hanoi University of Mining and Geology | http://www.humg.edu.vn |
| Hanoi University of Communication and Transportation | http://www.uct.edu.vn |
| Hanoi University of Water Resources | http://www.hwru.edu.vn |
| Hanoi University of Architecture | http://www.hau.edu.vn |
| Hanoi University of Agriculture No. I | http://www.hau1.edu.vn |
| University of Forestry | http://www.vfu.edu.vn |
| Thai Nguyen University of Technology | http://www.dhktcn.edu.vn |
| Hai Phong Maritime University | http://www.vimaru.edu.vn |
| Hue University of Agriculture and Forestry | http://www.huaf.edu.vn/ |
| Da Nang University of Technology | http://www.dut.edu.vn |
| Nha Trang University of Technology | http://www.ntu.edu.vn |
| Ho Chi Minh City National University | http://www.hcmut.edu.vn |
| Ho Chi Minh City University of Architecture | http://www.hcmuarc.edu.vn |
| Ho Chi Minh City University of Industry | http://www.hui.edu.vn |
| Ho Chi Minh City University of Agriculture & Forestry | http://www.hcmuaf.edu.vn |
| Ho Chi Minh City University of Technical Education | http://www.hcmute.edu.vn |
| Can Tho University | http://www.ctu.edu.vn |

Having been organized around the monodiscipline university model, many universities provide a limited number of training programs. Because of limited funding, most newly established universities offer engineering or technology programs that do not require high investments, such as computer, software, electrical, or electronic engineering. The list below shows the undergraduate engineering programs offered by all 172 universities in Vietnam. These data were collected by the authors in late 2007 from an official document of the Ministry of Education and Training (MOET 2007a). The data reveal quite a small number of programs in areas such as aerospace engineering, nuclear/radiological engineering, and marine/ocean engineering.

| Programs | Quantity | %* |
|---|---|---|
| Aerospace Engineering | 2 | 1 |
| Agricultural Engineering(Forestry and Aquaculture) | 19 | 11 |
| Refrigeration/Heat Engineering | 8 | 5 |
| Architectural/Design Engineering | 19 | 11 |
| Automotive Engineering | 16 | 9 |
| Biological Engineering | 21 | 12 |
| Chemical/Petroleum Engineering | 14 | 8 |
| Civil/Construction Engineering | 36 | 21 |
| Computer/Software Engineering | 66 | 38 |
| Electrical/Electronics Engineering | 67 | 39 |
| Environmental Engineering | 28 | 16 |
| Food Engineering | 16 | 9 |
| Geological/Material Engineering | 6 | 4 |
| Industrial Engineering | 8 | 5 |

(continued)

(continued)

| Programs | Quantity | %* |
|---|---|---|
| Manufacturing Engineering | 12 | 7 |
| Marine/Ocean Engineering | 2 | 1 |
| Mechanical Engineering | 17 | 10 |
| Nuclear/Radiological Engineering | 2 | 1 |
| Telecommunication Engineering | 31 | 18 |

*$N = 172$

According to the Education Act (Vietnam National Assembly 2005, p. 14), curricula of all programs (from diploma to postgraduate) in Vietnamese higher education must follow the curriculum frameworks provided by the Ministry of Education and Training (MOET). These frameworks are developed by groups of professionals in each area and then validated by the Minister. The main purpose of utilizing these curriculum frameworks is to insure a minimum level of teaching and learning quality in all universities in Vietnam. Within these frameworks, the higher the subject level, the more freedom a university has in creating its own curricula. Of course, this task takes considerable time and resources due to the heavy workload involved in producing curricula frameworks for all programs.

## Historical Background of Quality Assurance in Vietnam's Higher Education

Quality assurance for Vietnamese higher education emerged as an issue in the late 1990s. In a paper introduced at the World Conference on Higher (MOET 1998, p. 6), the Vietnamese delegation presented the strategy for higher education development in Vietnam up to 2020.

> To complete the system of higher education organization and management with the aim of increasing management effectiveness of the central governing body to every university, building and completing the laws relating to higher education. To expand the right of self-management hand in hand with heightening the responsibilities of universities as regards their training, scientific research and productivity. To develop a system of assessment and supervision of the quality of higher education, and to control the quality of the teaching and training and classification of universities based on a single set of criteria.

The National Workshop on Quality Assurance in Higher Education was held in Dalat, Vietnam in 2000, and focused on defining quality in higher education and the steps necessary to improve the quality of Vietnamese institutions. More significantly, at the Vietnam National Conference on Higher Education held in October 2001, the Prime Minister of Vietnam stated:

> MOET has to define its role in educational management, and in the coming decade it needs to focus on the following missions: establishing and developing plans and strategies for the development of education; establishing mechanisms, policies, and approaches in management of courses and quality of education by monitoring, inspecting and accrediting (Phan 2001).

Also in 1990s, the first two Quality Assurance centers were established at the two biggest universities in Vietnam, Hanoi National University and Ho Chi Minh City National University. Recognizing the importance of Quality Assurance and accreditation for higher education, MOET established the Division of Educational Quality Accreditation in early 2002 and then upgraded its Department of Testing and Accreditation in July 2003. This department is in charge of designing testing and accreditation plans (mechanisms and procedures) for all educational institutions in Vietnam (from elementary schools to universities), and plays a key role in the accreditation process.

In 2001, the Prime Minister approved the Educational Development Plan 2001–2010 which requires MOET to immediately establish and implement the accreditation process at all levels and types of education (MOET 2006, p. 60). Following this plan, in 2004, MOET approved the Provisional Regulation on Higher Education Accreditation (MOET 2004) which authorizes accrediting activity. The regulation establishes a national quality accreditation system for all higher education institutions (colleges and universities at both undergraduate and postgraduate levels) as well as training institutions, either public or private. It should be noted that at present, accreditation is just at the institution level.

Together with this regulation, in 2004 MOET approved a Set of Standards for Accreditation for all higher education institutions. This set comprised ten standards covering 53 criteria to be used for institution level accreditation. While the criteria were described in a quantitative style, these standards were expressed at a general level as follows:

Standard 1: Mission and objectives of the university
Standard 2: Organization and management
Standard 3: Training programs
Standard 4: Training activities
Standard 5: Managerial staff, lecturers, and staff
Standard 6: Learners
Standard 7: Scientific research and technology development
Standard 8: International cooperation
Standard 9: Library, learning equipment, and other facilities
Standard 10: Finance and financial management

In support of the national movement in Quality Assurance, the first Education Act (Vietnam National Assembly 2005) has a separate article on accreditation of educational quality that is defined as "the crucial means for identifying the attainment of educational objectives, curricula, and contents from schools and institutions" (p. 5).

As required by the Education Act, all universities in Vietnam have started to establish their own units of Quality Assurance and/or accreditation. In 2005, 20 universities around Vietnam (including 18 public and 2 private) were selected by MOET for a pilot accreditation process. This pilot process was part of the Higher Education Project funded by the World Bank with consultation by international professionals from the US and the Netherlands.

Each of those universities had around 6 months for a self-assessment process that was based on the above set of standards. The result was a self-assessment report that was sent to MOET. An external review team was then formed by MOET to validate each university's report and make its own recommendations to MOET as part of a final accreditation report.

Based on the experience gained from this pilot accreditation process as well as further studies on accreditation processes around the world, MOET approved the Regulation on Standards for University Accreditation (MOET 2007b) and three new Sets of Standards for Accreditation. These sets are separate for universities, colleges, and vocational schools, and used for institution level accreditation.

For universities, the new set has ten standards covering 61 criteria that are framed in a much more qualitative style than those in the previous set. Compared with the provisional regulation, the concept of "educational quality" has been redefined from "the satisfaction of the objectives determined by an institution" (MOET 2004) to "the satisfaction of the educational objectives determined by an institution, by the Education Act; of the human resource training needs for the improvement of the socio-economy of a province and the whole nation" (MOET 2007b).

Following the pilot accreditation of the first 20 universities, more universities began the accreditation process in 2007. However, the lack of Quality Assurance experts at institutions has forced MOET to slow down the process. In addition, the self-assessment process is a real burden for most institutions since they do not receive funding from the Higher Education Project.

## Program Accreditation Initiative

In August 2007, MOET drafted a Regulation on program accreditation procedures and cycles for universities, colleges, and vocational schools (MOET 2007c) for comments from institutions. According to this regulation, a program within an institution can only apply for accreditation if that institution has been accredited. Criteria for program accreditation in each area will be provided by MOET. The program accreditation procedure is similar to the institutional accreditation process. In November 2007, MOET approved the first official regulation on program accreditation for elementary teacher training programs which includes a set of seven standards and 37 criteria. These were finalized after an experimental period of implementing self-assessment and external visits for the first ten colleges of education that was begun in 2001 (MOET 2007d).

At the institution level, Ho Chi Minh City National University (VNU-HCM) is the first one to take the initiative in organizing program accreditation for its affiliated universities: University of Technology, University of Natural Sciences, University of Social Sciences and Humanities, University of Information Technology, International University, and College of Economics. According to its regulation issued in September 2007 (VNU-HCM 2007), programs of member universities can be audited/accredited based on criteria established by the Association of Southeast Asian Nations (ASEAN) University Network (AUN) or an international accreditation organization such as ABET. This has set the stage for the accreditation of engineering programs.

## Overview of Current Quality Assurance Scheme in Vietnam's Engineering Education

As mentioned above, all higher education programs in Vietnam must follow the curriculum frameworks designed by MOET in order to maintain a minimum standard of quality across institutions. Regarding engineering education, in November 2007, MOET officially approved the first curriculum frameworks for 14 undergraduate programs as follows (MOET 2007e).

- Heat and Refrigeration Engineering
- Electrical/Electronics Engineering
- Control and Automation Engineering
- Metallurgy Engineering
- Aerospace Engineering
- Food Engineering
- Textile and Garment Engineering
- Mining Engineering
- Geological and Mapping Engineering
- Petroleum Engineering
- Construction Material Engineering
- Construction Machine Engineering
- Water Supply and Drainage Engineering
- Bridge and Road Engineering

These frameworks are used for 5- or 4.5-year programs in which a maximum 260 credits are divided into two blocks: compulsory credits and elective credits (elective credits vary from about 50 to 120 depending on the program). While the compulsory credits must follow the syllabus descriptions within the MOET approved frameworks, institutions are free to design the elective ones.

Thus far the accreditation process in higher education in Vietnam mainly is at the institution level, while program accreditation is only being piloted in some teacher training programs. Ho Chi Minh City National University is the first university in Vietnam to initiate program accreditation by approving a Quality Assurance plan for the period 2007–2010. This plan comprises of two main activities: auditing academic programs based on AUN criteria and accrediting affiliated institutions based on MOET criteria.

## Initial Efforts in Applying International Accreditation Standards: Efforts in Applying ASEAN University Network (AUN) Criteria

The ASEAN University Network (AUN) was founded in November 1995 by Association of Southeast Asian Nations (ASEAN) member countries including 13 universities and increased to 20 member universities by 1999. Ho Chi Minh City

National University and Hanoi National University are the two members from Vietnam. In order to enhance Quality Assurance among the network institutions, AUN established AUN-QA (Quality Assurance) in 1998. After several workshops and conferences, AUN-QA Guidelines and a Manual for the Implementation of the Guidelines were published. The guidelines and manual provide a road map for what the AUN-QA calls "the journey to uplift the quality of higher education in ASEAN universities" (AUN-QA 2006, p. 7). Together they describe standards and criteria for accrediting member institutions and programs.

At the program level, the guidelines provide 17 standards with 53 criteria that can be used as a common framework for auditing or accrediting undergraduate programs. The topics of the standards are as follows:

- Goals and objectives; expected learning outcomes
- Program content
- Program specification
- Program organization
- Didactic concept/teaching/learning strategy
- Student assessment
- Staff quality
- Quality of the support staff
- Student quality
- Student advice and support
- Facilities and infrastructure
- Quality assurance
- Student evaluation
- Curriculum design
- Staff development activities
- Feedback stakeholders
- Output

As the official members of AUN, Ho Chi Minh City National University and Hanoi National University are planning to use the AUN-QA standards and criteria for auditing and accrediting their undergraduate programs. This will take place prior to their applying for the AUN Quality Label.

In the Ho Chi Minh City National University's Quality Assurance Plan for the period 2007–2010 (VUN-HCM 2007), the second objective includes:

By the end of the academic year 2007–2008, all affiliated institutions have to finalize their own self-assessment processes for the programs which aim to access international standards.

By the end of the academic year 2008–2009, all the above programs are audited based on AUN-QA criteria.

At the 2007 meeting of Quality Assurance units of AUN's members the first program accreditation plan at Ho Chi Minh City National University and Hanoi National University was approved (Vu 2008, p. 14). Therefore, it is expected that in 2009, engineering education, computer, and electronics – communication programs will be among the first group to be audited based on AUN criteria.

These programs will be able to take advantage of expertise available through AUN/SEED-Net (Southeast Asian Engineering Education Development Network). This network was established in 2001 as an autonomous subnetwork of the AUN. It is intended to help improve human resources and training quality in engineering education. Hanoi University of Technology and University of Technology – Ho Chi Minh City National University are presently members of the network.

## Efforts in Applying ABET Criteria

The Boeing Company, in cooperation with the ABET, Inc., sponsored a series of workshops in 2006 to familiarize Vietnamese universities with accreditation requirements for engineering education. The workshops sought to elucidate the objectives and processes of accreditation and foster curriculum planning that would enhance engineering programs in Vietnam. The workshops were organized at University of Technology – Ho Chi Minh City National University, University of Da Nang, and Hanoi University of Technology.

After the workshops, the participants examined their own engineering programs in depth, consulted with industry and government partners on educational objectives, revised curricula, and began a coordinated plan to improve the quality of engineering education offered to their students. The Boeing Company is working with ABET and Vietnamese universities to create an action plan for future implementation of the ABET criteria.

The University of Technology – Ho Chi Minh City National University is the first institution to have a detailed plan for applying ABET criteria for self-assessment and accreditation of its undergraduate engineering programs. According to this plan, there are three programs to be accredited by ABET by 2010 (computer engineering, chemistry engineering, and electrical and electronics engineering) with the other four by 2015 (civil engineering, environmental engineering, mechanical engineering, and industrial management) (Truong 2007).

## Conclusions: Recommendations for the Future

The following five recommendations are based on the experience of developing and implementing Quality Assurance schemes in Vietnam over the last 10 years. The first and most important recommendation is that consideration be given to establishing professional accreditation organizations which can help MOET in the accreditation process. Second, professional manuals and guidelines should be developed to direct and facilitate the implementation of the accreditation standards and criteria.

Next, future sets of standards and criteria need to focus on encouraging universities to fulfill their missions, not just to meet the minimum standards.

Fourth, they need to focus on the social accountability, and the transparency of higher education institutions. And finally, self-assessment, peer review, and student participation should have more emphasis in the process of accreditation (Nguyen 2003; Le 2007)

As discussed above, some undergraduate engineering programs at Vietnamese institutions are becoming dominant, such as computer and software engineering, electrical and electronics engineering, and telecommunication engineering. With the limited funding from the Government and institutions, it is difficult for all of these programs to provide quality instruction. A national accreditation plan for these areas should be a priority (Hayden and Lam 2006) in order to "protect students from poor quality programs, which are likely to occur in a rapidly expanding higher education system that is simultaneously diversifying" (Nguyen 2003, p. 256).

As identified by Lenn (2004), "accreditation is a proven means for improving higher education nationally as well as regionally" (p. 27), and "as regional identity becomes stronger and as international standards of quality become increasingly important to systems of higher education, Quality Assurance systems do not have to remain only national in nature" (p. 31). Accrediting undergraduate programs in general, and engineering in particular, is not only necessary for improving the quality of national programs themselves but also a crucial step toward integrating Vietnamese engineering education into regional and international higher education. Therefore, the implementation of regional and international accreditation standards such as those devised by AUN-QA or ABET, with modifications to make them applicable in the Vietnamese context, should be considered as a national long-term strategy.

Within AUN, such countries as Singapore and Thailand have had much experience in Quality Assurance and accreditation. The establishment of a regional pool of experienced external reviewers in the area of engineering education can help to decrease the gap between Vietnamese engineering programs and others within the region. Such a collaboration can "provide an international perspective of quality" (Lenn 2004, p. 31) to Vietnamese institutions.

## References

AUN-QA. (2006). *Manual for the implementation of the guidelines*. ASEAN University Network – Quality Assurance.

Hayden, M., & Lam, Q. T. (2006). A 2020 vision for higher education in Vietnam. *International Higher Education, 44*, 11–13.

Le, V. H. (2007). Vai tro cua sinh vien doi voi hoat dong quan tri va dam bao chat luong o dai hoc trong boi canh hoi nhap quoc te [The role of students in university governance and quality assurance in global integration context]. *Proceedings of the Conference on Higher education: opportunities and challenges in integrated and global context, Dalat, April 13–15.*

Lenn, P. M. (2004). *Strengthening World Bank support for quality assurance and accreditation in higher education in East Asia and the Pacific*. Working paper series on quality assurance and accreditation in higher education in East Asia and the Pacific (Paper No. 2004-6), World Bank.

MOET. (1998). *Vietnam*. Paper presented at the world conference on "Higher Education in the Twenty First Century" on October 5–9, Paris, France.

MOET. (2004). *Quy dinh tam thoi ve kiem dinh chat luong truong dai hoc, ban hanh kem theo QD so 38/2004/QD-BGD&DT* [Provisional regulation on university accreditation, issued as the attachment of the Decision No. 38/2004/QD-BGD&DT].

MOET. (2006). *Tong quan ve he thong dam bao chat luong giao duc dai hoc* [Overview on accreditation systems in higher education]. Hanoi: Department of Testing and Accreditation.

MOET. (2007a). *Nhung dieu can biet ve tuyen sinh dai hoc va cao dang nam 2007* [Substantial information on university and college entrance examinations]. Hanoi: Giao Duc Publisher.

MOET. (2007b). *Quy dinh ve tieu chuan danh gia chat luong truong dai hoc, ban hanh theo QD so 65/2007/QD-BGD&DT* [Regulation on standards for university accreditation, issued as the attachment of the Decision No. 65/2007/QD-BGD&DT].

MOET. (2007c). *Du thao Quy dinh ve qui trinh va chu ky kiem dinh chat luong chuong trinh giao duc cua cac truong dai hoc, cao dang, va trung cao chuyen nghiep* [Drafted regulation on programme accreditation procedures and cycles for universities, colleges, and vocational schools].

MOET. (2007d). *Qui dinh ve tieu chuan danh gia chat luong chuong trinh giao duc nganh giao duc tieu hoc trinh do cao dang, ban hanh theo QD so 72/2007/QD-BGD&DT* [Regulation on standards for accreditating elementary teacher training programmes, issued as the attachment of the Decision No. 72/2007/QD-BGD&DT].

MOET. (2007e). *Bo chuong trinh khung giao duc dai hoc khoi nganh ky thuat trinh do dai hoc, ban hanh theo QD so 70 /2007/QD-BGD&DT* [Curriculum frameworks for undergraduate engineering programmes, issued as the attachment of the Decision No. 70/2007/QD-BGD&DT].

Nguyen, K. D. (2003). *International practices in quality assurance for higher education teaching and learning: Prospects and possibilities for Vietnam*. Unpublished PhD thesis, University of Melbourne.

Phan, V. K. (2001). *Chien luoc phat trien giao duc 2001–2010* [Educational development strategies 2001–2010]. Paper presented at the conference on "Higher Education" on October 1–3, Hanoi, Vietnam.

Truong, C. H. (2007). *Ke hoach phan dau xay dung cac nganh dao tao dat chuan kiem dinh cua ABET (trong Tai lieu Hoi nghi chat luong dao tao – DHQG TP. HCM)* [Plan on improving programme quality for ABET accreditation (in Materials of Workshop on educational quality – VUN-HCM)].

Vietnam National Assembly. (2005). *Luat Giao Duc* [Education Act]. Hanoi: Giao Duc Publisher.

VNU-HCM. (2007). *Qui dinh tam thoi ve kiem toan va kiem dinh chat luong giao duc tai Dai hoc Quoc gia TP. Ho Chi Minh, ban hanh theo QD so 875/QD-DHQG-KT&DG* [Provisional regulation on auditing and accreditation at VUN-HCM, issued as the attachment of the Decision No. 875/QD-DHQG-KT&DG].

Vu, T. P. A. (2008). *Bao cao ve Hoi thao chat luong giao duc dai hoc va cuoc hop lan thu 8 cua AUN-QA (trong Tai lieu Hoi nghi tap huan dam bao chat luong trong giao duc dai hoc)* [Report on the Conference on higher education quality and the 8th meeting of AUN-QA in Malaysia (in Materials for the Workshop on quality assurance in higher education)], VUN-HCM.

VUN-HCM. (2007). *Ke hoach dam bao chat luong tai Dai hoc Quoc gia TP Ho Chi Minh giai doan 2007 – 2010, ban hanh theo QD so 1158/QD-DHQG-KT&DG* [Quality assurance plan of VUN-HCM in 2007–2010, issued as the attachment of the Decision No. 1158/QD-DHQG-KT&DG].

# Quality Assurance for the Engineering Paraprofessional in Thailand

**Kalayanee Jitgarun, Paiboon Kiattikomol, and Anuvat Tongsakul**

**Abstract** This chapter focuses on Quality Assurance in a context of industrial education and engineering paraprofessionals in Thailand. Presently, there are no defined Quality Assurance standards or competencies for these individuals in Thailand. In this chapter, we review the development of quality standards for vocational education in this country. We then review four different perspectives on Quality Assurance and competencies for engineering education. We discuss these in relation to the engineering paraprofessional in Thailand. We conclude with some suggestions for future directions for the development of competencies for engineering paraprofessionals in Thailand and developing countries. We suggest that these competencies must embrace lifelong and flexible learning but also that there must be an emphasis on soft skills related to attitudes and ethics. Distance education will also play an important role in Quality Assurance for the engineering paraprofessional.

## Introduction

The Faculty of Industrial Education and Technology at King Mongkut's University of Technology in Thailand, like others of its kind in the country, is responsible for technical (vocational) teacher training. The Faculty aims to produce trainers of technicians and to carry out research and development related to electrical, mechanical, civil, and production technology education. Instructors of these programs are working in two disciplines: vocational education and paraprofessional engineering yet Quality Assurance of these programs presently falls under the standards and auspices of vocational education only. In this chapter, we outline some possibilities for a new approach to Quality Assurance and competencies for engineering paraprofes-

K. Jitgarun (✉)
Electrical Technology Education Department, Faculty of Industrial Education and Technology, King Mongkut's University of Technology Thonburi, Bangkok, Thailand
e-mail: kalayanee.jit@kmutt.ac.th

A.S. Patil and P.J. Gray (eds.), *Engineering Education Quality Assurance: A Global Perspective*,
DOI 10.1007/978-1-4419-0555-0_8, 

sionals in this country. We accomplish this by drawing on Quality Assurance in the field of vocational education in Thailand and engineering education in higher education from other countries. We begin with an overview of Quality Assurance in education in Thailand in general then follow with a brief history of Quality Assurance in vocational education in this country. Next, we synthesize four different perspectives on the competencies for engineering education. Finally, we suggest some possible directions for the development of Quality Assurance standards for engineering paraprofessionals in Thailand.

## Education in Thailand and the Need for Quality Assurance

Improving national competitiveness is especially important in Thailand, which is not a country with a highly competitive labor force, due in large part to a weak educational infrastructure. In fact, according to the 2004 World Competitiveness Yearbook (International Institute for Management Development 2004) in which the world's 60 leading economic countries were studied, Thailand has a relatively low educational competitiveness rating compared with other developed countries around the world. For example, test scores of 15-year-old school children for 2000 and 2006, provided by the Organization for Economic Co-operation and Development Programme for International Student Assessment (OECD PISA), showed that, out of 57 countries, Thailand ranked 44th in science, 41st in reading literacy, and 43rd in mathematics.

As a result of such low rankings and test scores over the years, in 1999, Thailand began its educational reform efforts and initiated Quality Assurance in education. Quality Assurance is a means of ensuring that suppliers (higher education institutions themselves, including their administrators, funding sources, instructors, and staff) and customers (students, parents, employers, and society in general) are satisfied with both the quality and consistency of higher education. In addition, in areas such as engineering education, Quality Assurance can help ensure that a country is more competitive by providing graduates with requisite knowledge, skills, and dispositions. Furthermore, Quality Assurance can guide improvements in the quality of educational institutions by helping them identify their strengths and weaknesses. It is assumed that if its educational institutions adopt a system of Quality Assurance, based on international standards, then Thailand will become more competitive. Based on this assumption Thailand's National Education (NEA) Act (Ministry of Education 1999) states:

> ...There shall be a system of education Quality Assurance to ensure improvement of educational quality and standards at all levels. Such a system shall be comprised of both internal and external Quality Assurance...The system, criteria, and methods for Quality Assurance shall be as stipulated in the ministerial regulations.

In order to effectively institute Quality Assurance in higher education there must be an organization that sets the criteria and the methods to be used. In this regard, Section 49 of the NEA states that:

> An Office for National Education Standards and Quality Assessment shall be established as a public organization, responsible for development of criteria and methods of external evaluation, conducting evaluation of educational achievements in order to assess the quality of institutions, bearing in mind the objectives and principles and guidelines for each level of education as stipulated in this Act...

As a result, the Office for National Education Standards and Quality Assessment (ONESQA) was established in 2000 to evaluate the quality and certify educational standards in all higher education, vocational education, and fundamental education institutions in Thailand. Assessment results are intended to be used by educational institutions as a guide for achieving their ultimate goal which is the best preparation of learners in order to advance Thailand's global and regional competitiveness (National Education Standards and Quality Assessment 2007b). Section 49 also stipulated that all educational institutions shall undergo an external quality evaluation at least once every 5 years and that the results of the evaluation shall be submitted to the relevant agencies and made available to the general public (Ministry of Education 1999). The objectives of external Quality Assurance are as follows (The Office for National Education Standards and Quality Assessment, Public Organization 2007a):

1. To examine and verify the real status of work done by educational institutes and to assess the educational quality as specified by the educational standards.
2. To acquire the information showing the strengths and weaknesses of educational institutions as well as the origin of the problems and the conditions for success.
3. To help suggest improvements and develop the educational quality for educational institutions and their affiliations.
4. To empower educational institutions to develop and assess their internal quality continuously.
5. To report the results of quality assessment and educational standards of educational institutions to organizations concerned as well as the general public.

Given the roles and the responsibilities of ONESQA, it is important that it be a public organization under government supervision, i.e., neither a government organization nor a state enterprise. This makes it independent so that it can effectively decide its own administration, management, and finance. Moreover, such independence ensures its impartiality and allows it to form judgments without any pressure from any other organizations. As a result, its decisions will be unbiased and provide an appropriate check and balance system for the certification of higher education quality.

## Quality Assurance and Vocational Education in Thailand

Quality Assurance of educational institutions in all levels, including the vocational education level, begins with the development of an internal Quality Assurance plan that is intended to guide a process of self-assessment and reporting. Ideally, this internal Quality Assurance process may also be used for external Quality Assurance review done by ONESQA. Hence, both external and internal quality assessment should be harmonized so as to move an institution in the same direction.

As for the External Quality Assurance for vocational education the expected outcomes are as follows (The Office for National Education Standards and Quality Assessment, Public Organization 2007a):

1. Educational institutions will engage in a process of continuous quality improvement so that they eventually meet international standards.
2. There will be efficient resource management in educational institutions.
3. School-Based Management (SBM) will be implemented on the basis of the decentralization of the decision-making authority to the school in terms of budget, curriculum, and personnel decisions (Oswald 1995).
4. Students, parents, employers, and general public will have the information needed to make systematic judgments of institutional and/or program quality and, therefore, also to make appropriate decisions.
5. Educational institutions, educational administration offices, and government will have accurate and systematically collected information needed to formulate policies and plans and to manage education.

More specifically, Section 49 of the NEA sets the following criteria for educational management in vocational education regarding Quality Assurance (The Office for National Education Standards and Quality Assessment, Public Organization 2007a):

1. It must involve lifelong learning to produce and develop capacity.
2. It must develop the contents and the body of knowledge.
3. It must formulate the curriculum framework of Vocational Standards, Professional Standards, Occupational Standards, and Vocational Qualification, and there must be Quality Assurance for all vocational education levels (Vocational Certificate and Higher Vocational Certificate) in every vocational field.
4. It must have principles to promote teachers' professional standards continuously.
5. There should be public sectors, enterprises, institutions, and other societies in collaboration with vocational education and training.
6. Resources from all sites should be mobilized for educational management.
7. There should be collaboration among people, families, communities, local administration units, private sector, public organization, vocational institutions, religions, and other societies.

From 2001 to 2005, The Office for National Education Standards and Quality Assessment (ONESQA) conducted the first round of external Quality Assurance reviews of 765 public and private sectors institutions (The Office for National Education Standards and Quality Assessment, Public Organization 2007b). Round II will be conducted between 2006 and 2010.

The main focus of Round I assessments was institutional strengths and weaknesses and suggestions for educational development. These reviews were conducted under the clear, impartial, and just principles called *amicable assessment*, which was intended to support and encourage educational institutions to have a positive attitude toward external Quality Assurance, to trust outside auditors/assessors as colleagues, and to embrace the principles of authentic assessment. Amicable assessment was considered an appropriate approach given that Round I was involved in beginning the development of a culture of assessment.

The results of the assessment were not used to determine institutional success or failure but to help develop educational institutions and to meet national standards (see Standards I–VIII which will be discussed later). In addition to helping them meet standards, the reviews were intended to guide institutions in adopting such practices as self-assessment, institutional development, realistic assessment, standard-criterion evaluation, participatory assessment, and qualitative assessment. Moreover, ONESQA provided supporting resources for educational institutions so that assessment could be a continuous process and so that external Quality Assurance helped the institutions achieve their specified mission (The Office for National Education Standards and Quality Assessment, Public Organization 2007b).

The first round of external vocational education Quality Assurance review had a great influence on vocational education since the results, along with the key issues and suggestions were reviewed and submitted to the government. This made all stakeholders realize the importance of the Quality Assurance and of the quality of education. In addition, the results led to many modifications of Quality Assurance procedures to be undertaken in the second round. For example, each educational institute developed and revised its own internal Quality Assurance processes in order to better serve the purpose of external Quality Assurance. In addition, the government provided sufficient, reliable information to support the educational institutes that needed improvement in quality. As a result of the first round of Quality Assurance, the general public came to understand what was happening in Thai education better than it had in the past. Thus, the concept of Quality Assurance was well received by both the staff in educational institutions and the general public.

## Policy Suggestions

Suggestions for policies concerning quality development for vocational education needed by the government and the Office of the Private Vocational Education Commission (OPVEC) include the followings (2007c):

- Formulate a vocational education act.
- Establish an institution for vocational qualification.
- Designate courses for further study and courses for the labor market.
- Enhance teacher quantity and quality.
- Let the private sector play a role in vocational education to reduce the governmental burden.
- Vocational education should vary in accordance with the fields offered by educational institutions.
- Instruction should be based on best practices with intensive coursework and practical experience.

The suggestions concerning the instructional quality development that should be fulfilled immediately by educational institutions are as follows (Office of the National Education Standards and Quality Assessment, Public Organization 2007c):

- Educational institutions should collaborate with one another.
- There should be vocational guidance for learners.
- The ideas of SBM should be promoted.
- Instructor knowledge should be upgraded.
- Students should be supported in their development of English language abilities.

External Quality Assurance by ONESQA for vocational institutes in Round II (Office of the National Education Standards and Quality Assessment, Public Organization 2007a) is intended to certify the level of achievement of the Quality Assurance standards and to indicate strengths, weaknesses, opportunities, and threats as well as to give suggestions concerning the direction of development in the future.

Reports from educational institutions showed that most of the indicators used in the first round were unclear, resulting in misunderstanding and mistakes. Therefore, the following standards and indicators were revised in order to better correspond to the context of education (The Office for National Education Standards and Quality Assessment, Public Organization 2007c).

Standard I: The graduates are composed of three indicators as in: 1.1 the percentage of graduates who could pass the vocational standards, 1.2 the percentage of graduates who were employed within 1 year including those who were self-employed, and 1.3 the satisfaction levels of entrepreneurs/offices' supervisors of the graduates.

Standard II: Learning consisted of four indicators as in: 2.1 the number of hours for practical courses throughout the program, 2.2 the ratio of tools, equipment, devices, and learning materials utilized with cost–benefit, 2.3 the satisfaction levels of instructors, students toward tools, equipment, devices, and learning materials utilization, and 2.4 the number of hours for students' on-the-job training at enterprises with Memorandum of Understanding (MOU) throughout the program.

Standard III: Learning supports consisted of seven indicators as in: 3.1 the ratio of students and tenured instructors, 3.2 the operating cost per student, 3.4 the percentage of budget allocated for practical materials compared with the operating cost, 3.5 the ratio of tenured professional instructors and students in each field/vocational areas as well as teaching methods complies with National Education Act of B.E. 2542 (1999), 3.6 the overall expenses for Resources Center, and 3.7 the number of man-hours of experts/professionals from the business sector or local wisdom.

Standard IV: Research and creative work consisted of three indicators as in: 4.1 the number of innovations, projects, applied research/action researches, and academic articles of instructors/students, 4.2 the number of innovations, projects, applied research/action researches which could be utilized in teaching and learning or applied in business, industries, or the development of local community as well as the country, and 4.3 the amount of money granted for innovations, projects, applied research/action researches, and the academic works of all instructors.

Standard V: Academic services were composed of two indicators as in: 5.1 the number of activities/projects offered to communities/societies and 5.2 the percentage of the whole budget allocated to activities and projects offered to communities/societies each year.

Standard VI: Arts and cultural supports consisted of two indicators as in: 6.1 the number of activities and students who attended the activities and 6.2 the percentage of the whole budget allocated for student activities.

Standard VII: Administration and management consisted of seven indicators as in: 7.1 the administrators' leadership with good governance, 7.2 the percentage of salary for all staff as compared with the operating cost, 7.3 the ratio of nonacademic staff per student, 7.4 the percentage of expenses for central administration and management as compared with the operating cost, 7.5 the value of depreciating cost per student, 7.6 the budget allocated for maintenance per student, and 7.7 the percentage of the remaining budget as compared with the operating cost.

Standard VIII: Internal Quality Assurance consisted of two indicators as in: 8.1 the continuous systems and mechanism in internal Quality Assurance and 8.2 the effectiveness of internal Quality Assurance.

As for Round II, the main focus of these standards and criteria are output of vocational level education based on the internal quality procedure, which focuses on inputs and processes. Taken together the internal and external procedures make it clear that educational institutes should prepare annual internal Self-Assessment Reports to connect with external Quality Assurance requirements. Therefore, instead of eight, there are now six standards for external Quality Assurance with 25 indicators for vocational education (The Office for National Education Standards and Quality Assessment, Public Organization 2007a).

Standard I: Internal Quality Assurance consists of two indicators as in: 1.1 the systems and mechanism in internal Quality Assurance to develop Quality Assurance continuously and 1.2 the effectiveness of internal Quality Assurance.

Standard II: Quality of graduates is composed of four indicators as in: 2.1 the percentage of graduates who could pass the vocational standard criteria, 2.2 the learning achievement complied with the period of graduation criteria, 2.3 the percentage of employment within 1 year including those who were self-employed, and 2.4 the satisfaction levels of entrepreneurs/offices' supervisors of the graduates.

Standard III: Teaching and learning vocational education consists of ten indicators as in: 3.1 the development of competency-based curriculum which focuses on on-the-job training in order to empower students' abilities at international as well as dual levels, 3.2 the institutions should have various learning activities and systems which focuses on practicing with real professional skills, 3.3 the ratio of professional instructors per students in each program, 3.4 the number of man-hour of experts/professionals from business sector or local wisdom to be invited to give lecture to students in each field, 3.5 the satisfaction levels of students on the quality of instructors' teaching, 3.6 the percentage of budget allocated for practical materials compared with the operating cost in each field, 3.7 the readiness of Resource Center, 3.8 the sufficiency and currency of the equipment, tools, devices, and learning materials in each field, 3.9 the number of students activities and projects for students' development, and 3.10 effectiveness of students' activities both in academic, moral ethical aspects.

Standard IV: Innovations and body of knowledge for instructors and students consist of three indicators as in: 4.1 the number of innovations, inventions, action research of instructors and students to develop students' learning, 4.2 the number of inventions, innovations which are awarded from contests, national dissemination, and/or professional utilization, and 4.3 the percentage of the budget allocated to support the body of knowledge developed for instructors and students including the budget received from outside as compared with the operating cost.

Standard V: Academic services to communities and societies consist of two indicators as in: 5.1 the number of activities/projects offered to communities and societies and 5.2 the effectiveness of academic activities offered to communities and societies.

Standard VI: Administration and management consist of six indicators as in: 6.1 administrators at all levels having vision and leadership, the administrative plans formulated by collaboration with vocational education community and the responsibility to the success of work, 6.2 the database of institutions inside the institute to administer and manage, 6.3 the number of instructors who were trained in the teaching of their professional courses as well as the process of learning activities, 6.4 the development of work/project relevant to the strategic plan as well as emphasizing the cooperation of networking and vocational education community to utilize the resources together including the collaboration with enterprises to enhance educational management and/or dual program promotion.

For the second external Quality Assurance round, the NEA, Section 51 clarifies that when results of Quality Assurance indicate that an institution has not reached the required standards then *recommendations for corrective measures* will be provided to the institution by the Office for National Education Standards and Quality Assessment. If these measures are not acted upon within a given time, the Office is directed to submit reports to the Commission for Basic Education or the Commission for Higher Education so that the necessary remedial action can be taken.

Based on the experience gained from external Quality Assurance Round I, ONESQA suggested the following policies to develop educational quality and management standards in Round II (The Office for National Education Standards and Quality Assessment, Public Organization 2007b):

1. In order to change the values of vocational education, institutions charged with certifying Vocational Qualification should be established immediately. They should be able to make judgment on monthly payment based on skills and abilities of graduates instead of their certificates and diplomas.
2. Rules and regulations on educational Quality Assurance stated by the Ministry should be revised so that there is uniformity of internal and external Quality Assurance. There should also be Quality Assurance processes concerning institutional research, evaluation, and the continuous development of vocational education quality.
3. Educational institutions should develop a learning resources center for the benefit of all students.
4. There should be a center responsible for regularly conducting follow-up studies of graduate success.

Moreover, ONESQA developed a system for Round II to certify the educational standards and a program for preparing meta-evaluators. These are professionals who are qualified to conduct evaluations and who have passed a test that indicates that they understand Quality Assurance assessment (The Office for National Education Standards and Quality Assessment, Public Organization 2007d). This approach aims to provide assessors who are impartial, reliable, and professional.

## Engineering Education Quality Assurance

In this section of our chapter, we provide a synthesis of Quality Assurance standards for four different contexts. From these four, we will derive a preliminary set of possible Quality Assurance standards for the education of engineering paraprofessionals in Thailand.

The first set of quality standards we reviewed (Mott et al. 2002) focused on Mechanical Engineering Technology, Manufacturing Engineering Technology, and Industrial Engineering Technology. The authors highlighted certain factors that influence Quality Assurance standards. These are energy independence and environmental issues; the impacts of globalization on industry; and an increase in use of computer technology in industry. All of these issues affect the requirements for education. The implications of these issues include an increased need for distance education to support lifelong learning. In addition, there will need to be assessment to ensure that outcomes are met by students and the assessment will also require the development of clear student learning objectives and measurement tools. Other needs of programs will include interdisciplinary projects that emphasize *teamwork, co-op, internship, industrial employment* to ensure that programs maintain *currency* and *relevance to industry needs*. With regards to the latter, the authors suggest that students be expected to write papers about how their industry experience helped improve their abilities and attitudes (Mott et al. 2002).

The constant changes and innovation in industries also has implication for education. The needs for global competitiveness, consumer demand for higher performance, better quality, more customization, as well as lower costs demand the use of *efficient* and *innovative* technologies (Mott et al. 2002). They cite the President of the Society of Manufacturing Engineers who argues that *rapid change is a continuum and not a sequence of discrete events, and that we must react creatively to this new paradigm* (Mott et al. 2002, p. 3). The authors explain that *these technical and nontechnical trends reflect the paradigm of constant transition... These trends also challenge educators to consider what competencies their graduates need to demonstrate* (Mott et al. 2002, p. 4). Furthermore, effective educational programs will need to ensure a balance between applied learning and theory so that students can meet their field's need for lifelong learning and creativity. They also argue in favor of activity-based approaches and problem-centered activities as well as less reliance on lecture methods. In summary, they propose the following competencies: a recognition of the need for and an ability to engage in lifelong learning; an ability to understand professional, ethical, and social responsibilities; respect for diversity and a knowledge of contemporary professional, societal, and global issues; and a commitment to quality, timeliness, and continuous improvement.

Many of these same arguments have been put forth by other researchers. For example, Rugarcia et al. (2000) argue that to understand how engineers should be trained, we must understand the conditions and characteristics of the society within which engineers will function. They outline components related to knowledge, skills, and attitudes that engineering education graduates will need to possess as a result of these characteristics. In terms of knowledge, they recommend a move away from specialty training toward more cross discipline knowledge and the development of lifelong learning abilities. In terms of skills, they identify seven categories as follows (Rugarcia et al. 2000, p. 6):

1. Independent, interdependent and lifetime learning skills
2. Problem solving, critical thinking, and creative thinking skills
3. Interpersonal and teamwork skills
4. Communication skills
5. Self-assessment skills
6. Integrative and global thinking skills
7. Change

Finally, in terms of attitudes and values, the authors propose that engineers be willing to participate, be concerned about environmental preservation, hold a commitment to quality and productivity, and be involved in service to others. According to these authors, engineers must make decisions that take into account "the social, ethical, and moral consequences of those decisions" (Rugarcia et al. 2000, p. 10). Similar skills have been identified by the Accreditation Board for Engineering and Technology. (See the discussion of ABET's (a)–(k), by Peterson, in chapter "Quality Assurance in the Preparation of Technical Professionals: The ABET Perspective.")

Kastenberg et al. (2006) outline a set of five competencies for Engineering Education in the twenty-first century. Engineers should maintain a high level of technical expertise; develop an historical perspective in order to understand the nature and role of contexts and paradigms; develop an understanding of systems and networks in order to see the world holistically/ecologically; develop *ethical know-how*; and develop leadership and entrepreneurship (S1H-26).

## Quality Assurance and the Engineering Paraprofessional in Thailand

The competencies required for the engineering paraprofessional will be similar to those required for engineers. Technical training will be important to ensure that paraprofessionals are equipped with the technical skills required by their field. Technical competencies will, however, need to be constantly upgraded since new technological changes in industry will drive changes in the discipline. Their technical training will need to include an ability to apply theoretical concepts from, for example, the fields of mathematics and science. Within the context of technical training, engineering paraprofessionals, like their professional counterparts, will need knowledge and skills in computer technology. Given that the technologies as well as developments in industry are constantly changing, lifelong learning must figure as a core competency. Thailand's Educational Act (Ministry of Education 1999) already recognizes the importance of lifelong learning. What may be difficult is the translation of a competency for lifelong learning into measurable outcomes. Lifelong learning, as others have noted, will need to rely on distance education. The provision of distance education will therefore need to be included as a priority standard in Quality Assurance for the engineering paraprofessional for example where accreditation of institutes is concerned.

Technical training at a distance will necessitate the reliance on new and emerging technologies that go beyond simply the transmission of static content. Technical training of the engineering paraprofessional will call for state-of-the-art online tools that support skills development and applied and problem-based learning. Distance education opportunities can also help meet needs and face challenges related to globalization and the internationalization of industry. It can facilitate exposure to international industry standards. The education of engineering paraprofessionals will need to continue to emphasize, not only the importance of communications skills, but also the importance of communicating in languages other than Thai (e.g., Japanese, Chinese, and English). Once again, distance education can provide opportunities to attain these skills.

Quality Assurance standards could monitor as well as drive the quality of distance education programs. Whether their learning is at a distance or face-to-face, it will need to reflect the complexity, interdependence, and evolution of knowledge in the field of the engineering paraprofessional. Quality Assurance standards may require continuous updating to ensure that they reflect this character of knowledge. The need for collaboration and teamwork can be met by both face-to-face and distance programs. Quality Assurance standards can ensure that curriculum outcomes include an emphasis on knowledge- and skill-sharing and that assessment procedures value teamwork.

The competencies that may be the most difficult to monitor by Quality Assurance standards are those related to soft skills. These include attitudes, ethical conduct, professional and social responsibility, and concern for the environment. Quality Assurance provides a means to ensure that institutions offering education to paraprofessionals include an emphasis on these soft skills in their mission statements, in hiring practices, and in curriculum outcomes. To ensure that soft skills are valued by all institutions offering education for the paraprofessional, accreditation boards may need to be given authority and responsibility to verify these standards. In Thailand, where there exists a relative uniform culture given that the country's inhabitants are more than 90% Buddhist, agreement on the attitudes and social responsibilities may not prove too difficult. In other developing countries with various religions such as in India, reaching agreement may present more challenges.

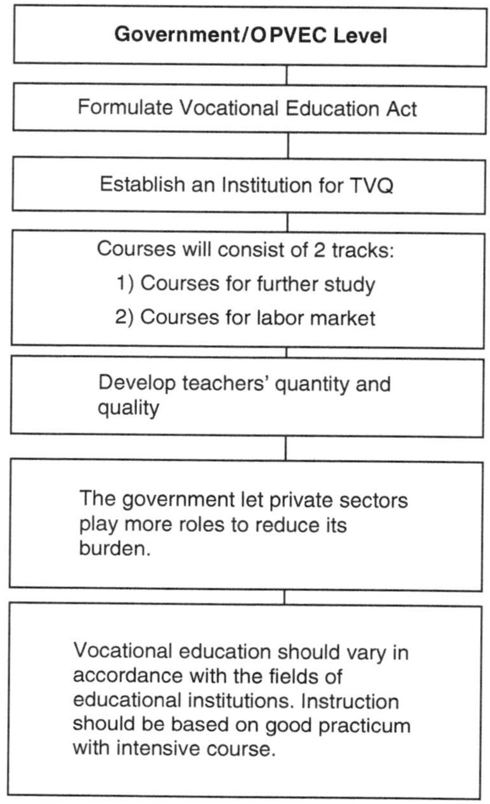


**Fig. 1** Suggestions in terms of policies to develop vocational education quality at the level of government/OPVEC

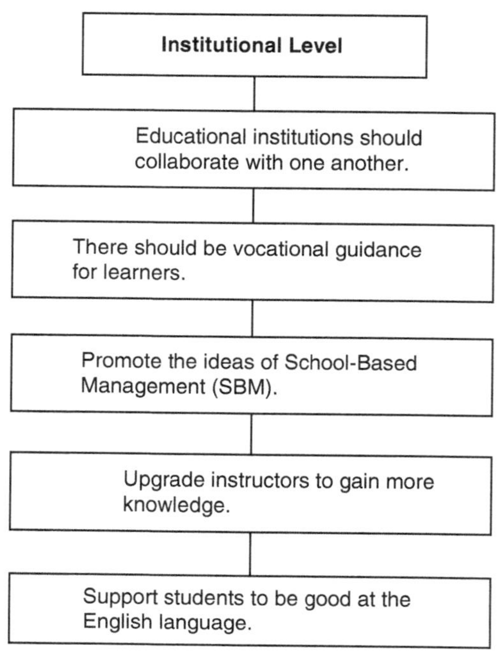


**Fig. 2** Suggestions in terms of policies to develop vocational education quality at the institutional level

## Conclusions

This chapter has provided a starting point for consideration of Quality Assurance for the engineering paraprofessional in Thailand. In many respects, the paraprofessional in Thailand is no different than his or her counterpart in other developing countries around the world since the competencies we outline for the paraprofessional reflect global trends and issues. Such globalization may in fact facilitate adoption of Quality Assurance for the engineering paraprofessional.

**Acknowledgments** The authors would like to extend a very special thank to Dr. Elizabeth Murphy, Associate Professor, Faculty of Education, Memorial University, Newfoundland and Labrador, Canada as well as Mr. Suthee Ploisawaschai, Assistant Researcher, Learning Institute, King Mongkut's University of Technology Thonburi for proofreading and editing this article.

## References

International Institute for Management Development. (2004). *World competitiveness yearbook 2004*. http://www.imd.ch/research/publications/wcy/index.cfm

Kastenberg, W., Hauser-Kastenberg, G. & Norris, D. (2006). *An approach to undergraduate engineering education for the 21st century*. 35th ASEE/IEEE Frontiers in Education Conference, San Diego, CA. Accessed 1 October 2008. http://ieeexplore.ieee.org/iel5/4116829/4116830/04117195.pdf?isnumber=4116830&prod=CNF&arnumber=4117195&arSt=23&ared=28&arAuthor=Kastenberg%2C+W.E.%3B+Hauser-Kastenberg%2C+G.H.%3B+Norris%2C+D

Ministry of Education. (1999). *National Education Act of B.E. 2542* (1999). Accessed 25 March 2008. http://www.moe.go.th/English/edu-act.htm

Mott, R., Neff, G., Stratton, M., & Summers, D. (2002). Future directions for mechanical, manufacturing, and industrial engineering technology programs. *Journal of Engineering Technology, 19*(1), 8–15.

Office of the National Education Standards and Quality Assessment (Public Organization). (2007a). *Handbook for External Quality Assurance to Certify Educational Standards for Vocational Education*, May, 2007 (pp. 4–5, 13–14, 18, 29–32). Bangkok: 21 Century Co. Ltd.

Office of the National Education Standards and Quality Assessment (Public Organization). (2007b). *ONESQA public report*, August, 2007 (pp. 91, 97, 98). Bangkok: Amarin Printing and Publishing (Public Organization).

Office of the National Education Standards and Quality Assessment (Public Organization). (2007c). *First external quality assurance report for vocational education between B.E. 2544–2548 (2001–2005) or Round I*, August, 2007 (pp. 61–62, 151). Bangkok: Amarin Printing and Publishing (Public Organization).

Office of the National Education Standards and Quality Assessment (Public Organization). (2007d). *ONESQA A public report. Daily Matichon*, Friday, 02 November, 2007 (pp. 24–25).

Oswald, L. J. (1995). School-based management. *ERIC Digest 99*. USA: College of Education, University of Oregon. Accessed 09 September 2008. http://eric.uoregon.edu/ publications/digests/digest099.html

Rugarcia, A., Felder, R. M., Woods, D. R., Stice, J. E. (2000). The future of engineering education-I: A vision for a new century. *Chemical Engineering Education, 34*(1), 16–25. Accessed 14 January 2009. http://www4.ncsu.edu/unity/lockers/users/f/felder/public/Papers/Quartet2.pdf.

# Quality Assurance in Higher Education in Chile: National and Engineering Dimensions

**Mario F. Letelier, Patricio V. Poblete, Rosario Carrasco, and Ximena Vargas**

**Abstract** This chapter describes the evolution of Quality Assurance in the Chilean higher education system. It emphasizes three relevant critical elements including the processes for conferring autonomy to private universities, the pilot project lead by the Ministry of Education, and the official status of the national higher education Quality Assurance system. There is also an analysis of the evolution of the Quality Assurance concept and related assessment mechanisms that during the past two decades have shifted in focus toward continuous improvement The accreditation process for engineering programs and a brief case-study that highlights the effects of accreditation at the School of Engineering of the country's oldest University are described as well. In addition, this chapter presents a brief overview of Quality Assurance experiences in other Latin American countries and concludes by pointing out further directions for development.

## Introduction

The design and implementation of a formal, official system of Quality Assurance in Chilean higher education started in 1990. The development of this system occurred in parallel with the country's social and economic improvement. Engineering program accreditation evolved during this period as well. This chapter discusses the stages in the evolution of the Chilean higher education Quality Assurance system.

---

M.F. Letelier (✉)
Centre for Research in Creativity and Higher Education, Universidad de Santiago de Chile, Casilla 10233, Santiago, Chile
e-mail: mario.letelier@usach.cl

A.S. Patil and P.J. Gray (eds.), *Engineering Education Quality Assurance: A Global Perspective*,
DOI 10.1007/978-1-4419-0555-0_9, 

## Quality Assurance at a National Level

Starting in 1981, the higher education landscape in Chile started to change, becoming increasingly heterogeneous and competitive. Up until then, there were eight universities in the country, plus some institutes, academies, and other institutions aimed at providing professional education. At that time the law regulating higher education in Chile established three kinds of civil institutions, namely, universities, professional institutes, and technical training centers. However, since 1981, all universities operate as nonprofit corporations whether they are privately funded or state supported, while institutes and centers have become for-profit societies. As a result, during the 1980s numerous institutions were created, some as a result of reshaping existing institutions, but there were also many new ones.

The development and implementation of accreditation procedures in the Chilean Higher Education System was made possible by a number of factors, a crucial one being the creation of the Higher Education Council (Consejo Superior de Educación – CSE) in 1990. The Council's role was to supervise and to grant authority to existing national private universities through accreditation-like procedures. After more than a decade of experience, the Consejo has rendered important lessons in higher education Quality Assurance. These have provided the foundations for the expansion of accreditation procedures to the traditional national public universities, as well as to other autonomous higher education institutions (Letelier 1997; Letelier et al. 2003).

In addition to the functions of this council, the interest in higher education evaluation in Chile is also explained due to the need to supply valid and reliable information to the public, especially students and parents, about the quality of different educational programs offered. Several factors including competition among institutions led them to promise certain student outcomes or to claim a particular quality of services. It was clear that there needed to be objective evaluations by external agencies of such promises and claims. Another reason that makes it desirable to have a Quality Assurance system is the current amount of Chile's international free-trade treaties with industrialized nations (Letelier and Carrasco 2004; Lemaitre 2002).

In 1999, the National Commission for Undergraduate Accreditation (Comisión Nacional de Acreditación de Pregrado – CNAP) began a voluntary program accreditation process. This initiative was intended to regulate the quality of higher education programs, and its mission was to implement experimental accreditation plans and to design a national Quality Assurance system. Its existence came at a time of unprecedented expansion of the Higher Education system in terms of enrollment, institutions, and programs.

From 1999 until 2002, the accreditation processes only included undergraduate programs of universities, professional institutes, and technical training centers. One of its main features was the voluntary participation of institutions. The assessment focused on the analysis of outcome profiles (perfiles de egreso) of undergraduate programs and the accomplishment of a set of assessment criteria. The assessment relied on a self-study process and an external audit by national and international experts.

Although for the first few years there was quite slow progress with relatively low participation, by December 2006 more than 560 programs had voluntarily engaged in the accreditation process. These programs represent almost 30% of the total undergraduate enrollment in Chile. By January 2007, CNAP had completed the assessment and accreditation process for almost 380 programs. Taking into account that the accreditation conferred to these programs could vary between nonaccredited and 7-year accreditation period, the results for 346 cases accredited by January 2007 were as follows: 15 did not receive accreditation, 53 were accredited for 2 years, 61 were accredited for 3 years, 80 were accredited for 4 years, 89 were accredited for 5 years, 26 were accredited for 6 years, and 22 were accredited for 7 years. Of the 1,660 technology programs eligible for accreditation (representing 23% of all undergraduate enrollments), 84 programs, which account for approximately 12% of this enrollment, have undergone the accreditation process (CNAP 2007).

At the beginning of 2003, CNAP announced the development of voluntary and experimental institutional accreditation plans in the form of academic audits. This decision represented a further step in terms of higher education Quality Assurance. The intent was to assess institutional management and teaching processes. There were four elective areas that could also be included in the audit, namely, research, graduate teaching, community engagement, and lifelong learning. During the first period, between May 2003 and April 2004, nine universities of the Board of Chancellors (Consejo de Rectores – CRUCH), three private autonomous universities, and two professional institutes decided to undergo institutional assessment and accreditation.

Once this experimental stage was completed, adjustments were made to the criteria and procedures in order to extend the accreditation process to other institutions. By January 2007, 66 institutions had completed this academic audit process. These 66 institutions enroll more than 70% of the undergraduates in the institutions eligible for accreditation.

Between 1999 and 2007, CNAP made significant progress in developing and implementing Chilean higher education Quality Assurance practices. The large number of undergraduate programs accredited as well as the number of institutional accreditation decisions made during this period are a sign of this progress. By January 2007, accreditation represented 100% of the undergraduate enrollment population of Board of Chancellors' universities, 72% of private universities, 83% of professional institutes, and 84% of technical training centers.

The experimental accreditation process lead by CNAP provided a variety of lessons related to human resource needs, institutional development, and academic teaching and learning processes. Concerning human resources, the experimental accreditation process highlighted the critical role of well-trained external assessment specialists familiar with different disciplinary areas and academic management. In addition, the accreditation revealed the importance of qualified internal professionals able to administer accreditation processes and, thus, provide the institutional know-how regarding self-regulation. In terms of institutional development, the experimental accreditation process increased the value given to assessment by pointing out those factors that have an impact on educational quality. Among the benefits to academic teaching and learning that resulted from the experimental accreditation were the better definition of outcomes profiles (which, nonetheless, still requires further elaboration) and the demonstrated value of incorporating alumni follow-up systems as a means of gathering input for improvement.

Beginning in January 2007, higher education Quality Assurance entered its second phase that is characterized by formal and obligatory accreditation processes (in contrast to the voluntary participation between 1999 and January 2007). In fact, this new phase had its origin in 2003, when CNAP presented a proposal to the Ministry of Education for the creation of a National Higher Education Quality Assurance System (Sistema Nacional de Aseguramiento de la Calidad – SINAC). This system has three main priorities: institutional accreditation, undergraduate and graduate program accreditation, and the authorization of accreditation agencies. This meant that the functions carried out by the National Commission of Graduate Accreditation (Comisión Nacional de Acreditación de Postgrado – CONAP), which was part of the National Commission of Technological and Scientific Research (Comisión Nacional de Investigación en Ciencia y Tecnología – CONICYT), were now absorbed by this new organization.

In November 2006, after the introduction of great number of modifications, the Higher Education Quality Assurance Law came into force. This law stipulated that the National Commission of Accreditation (Comisión Nacional de Acreditación – CNA) is to be in charge of the authorization and supervision of accreditation agencies and the implementation of a higher education information system.

The most innovative feature of the new system and law, besides the formal and obligatory character of the accreditation process, has been the creation of accreditation agencies. Since CNAP's accreditation experience demonstrated that one agency could not solely encompass the whole range of accreditation processes, the creation of multiple accreditation agencies was authorized. At present the agencies are:

- Acredita Ci, which is authorized to accredit undergraduate, graduate (master), and technical programs in agriculture (except veterinary), administration, technology, and commerce.
- AcreditAcción, authorized to accredit undergraduate, graduate, and technical programs related to education, health (except dentistry and medicine), technology, agriculture, administration, and commerce.
- Akredita, authorized to accredit undergraduate and technical programs in education, administration, commerce, health (with some exceptions), social sciences, and technology.
- Qualitas, authorized to accredit technical and undergraduate programs (except those of Pontificia Universidad Católica de Chile, Professional Centre DuocUC and Technical Training Centre DuocUC) in agriculture, social sciences, education, and technology.
- Agencia Acreditadora de Arquitectura, Arte y Diseño – AADSA, authorized to accredit technical, undergraduate, and graduate programs master in the areas of arts and architecture.

These agencies will be supervised by CNA and are responsible for accrediting undergraduate programs, master programs, and health professions in the areas and levels for which they have been designated.

## Evolution of Quality Assurance Concept in Chile

Quality assurance in Chile demands that institutions and programs declare their purposes, provide activities and resources to accomplish those purposes, produce corresponding outcomes and impacts, and make continuous improvements after evaluating the outcomes and impacts as shown in Fig. 1. The former assessment processes led by the CSE focused primarily on boxes 1 (purposes) and 2 (activities and resources) and weakly on boxes 3 (outcomes) and 5 (feedback and improvement). More recently, the accreditation system led by CNAP and now CNA takes into account all boxes, with special emphasis on outcomes, impacts, and Quality Assurance mechanisms. These are conceived as a composite of procedures, resources, and organizations that guaranty that the *quality cycle* of Fig. 1 is performed with increasing effectiveness.

So far, the concept of *Quality Assurance mechanisms* has proven to be a bit elusive for institutions that are less academically developed. In that sense, the concepts and their practical consequences represented in Fig. 1 provide a guide for quality improvement.

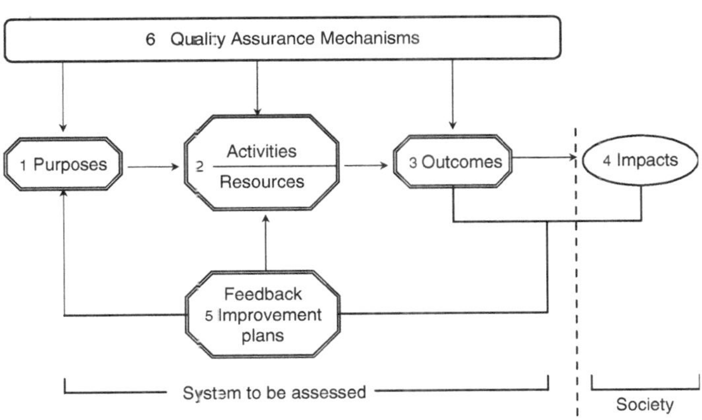


**Fig. 1** Quality assurance cycle in Chile

## Quality Assurance in Engineering

In Chile, engineering programs are called *Ingeniería Civil* (Civil Engineering) and usually last 5 or 6 years. The full name of a program is, for example, Chemical Civil Engineering or Mining Civil Engineering. At present there are a total of 168 of such engineering programs.

Accreditation of engineering programs started in 2002. For all programs, self-study and evaluation by external experts are conducted against nine criteria, namely, *purposes, integrity, organization, curricula, faculty, teaching effectiveness, outcomes, resources, and professional involvement*. Accreditation ranges from 2 to 7 years. Table 1 shows the accreditation duration for programs thus far accredited.

Based on the available data it is possible to describe the correlation between institutional and engineering program accreditation, as shown in Fig. 2, among state-supported universities. For every university, the plot indicates the relation between the average number of years of program accreditation (Civil Engineering)

**Table 1** Accreditation range of civil engineering programs

| Accreditation years | Number of programs |
|---|---|
| 0 | 1 |
| 2–3 | 20 |
| 4–5 | 28 |
| 6–7 | 22 |

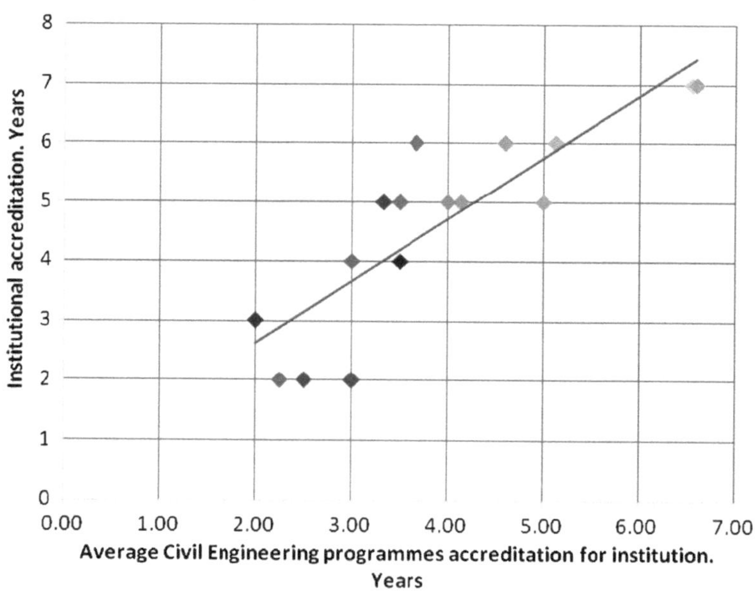


**Fig. 2** Correlation between institutional accreditation and program accreditation

and the number of years of institutional accreditation. The relatively high rate of 7-year accredited programs is due to the fact that the most prestigious universities entered the accreditation process much earlier. The rate should drop as more programs are subject to evaluation. The correlation shown in Fig. 2 is weak, which is expected to be so at this stage of the accreditation process.

Some important insights related to engineering education in Chile and that accreditations have revealed include (Letelier and Sandoval 2007):

- *Outcome profiles.* There is no common pattern for stating learning outcomes. This indicates a lack of involvement of programs.
- *Curricula.* In most cases curricula are not designed in close relation to the expected outcome profile. There usually appears to be a noticeable difference.
- *Fieldwork feedback.* Feedback proved to be almost nonexistent at the beginning. As accreditation progressed, this situation started to improve.
- *Program length.* Civil Engineering programs have a nominal duration of 6 years. However, students complete them, on average, in 7 or more years. This is due to several factors that ought to be corrected.
- *Student progression.* Graduation rates are low in general. In many cases, universities do not apply admission requirements that are consistent with expected student performance.
- *Quality Assurance cycle.* Most colleges of engineering have not developed a capacity for effectively maintaining continuous improvement. Rather, they occasionally act in reaction to accreditation.
- *Research.* Many universities state in their mission a strong commitment to research, which is not found to be the case in practice.

The following section describes the Quality Assurance approach and accreditation efforts in the University of Chile, which is the oldest and most prestigious university in the country.

## Case Study: Quality Assurance in the Engineering Programs at the University of Chile

The University of Chile, established in 1842, is the oldest in the country, and its Faculty of Physical and Mathematical Sciences, of which the School of Engineering and Science is part of, is one of the four original faculties of the university. It offers engineering programs in many different disciplines, and it is widely considered to be a leading school in Chile.

The University of Chile was not an early adopter when voluntary accreditation began in Chile, but as government grants began to make it a requirement to be accredited, preparations began to be made and finally, during 2006, the school participated in an accreditation process for the first time in its history. Because the new accreditation law had already been enacted and the old CNAP was winding down its operations, the process was actually a mixture of the old and the new: it was done according to the CNAP rules, but the entity in charge was one of the new authorized accreditation agencies. An additional complication arose from the fact that the school was launching a new curriculum for all its programs, starting in March 2007, but the accreditation was about the current (and soon to be old) programs.

The process began with the preparation of a number of self-study reports at the university, one per program being accredited. These reports were submitted in late 2006, and in March 2007 all of the peer-review teams appointed by the accreditation agency visited the university. In April 2007, CNAP published the results of this process and shared detailed evaluation reports with the university.

## Self-Study Reports

The reports included general information about the university, the faculty, and the program, including the program-learning objectives and the professional profile of graduates. Required statistical data about processes and human, infrastructure, and financial resources were obtained through the central information system that supports teaching and learning. Evidence of the effectiveness of different engineering programs was obtained through surveys directed to students, alumni, and employers, which were conducted by each department.

A characteristic of this school is that incoming students do not enroll directly into their chosen program. Instead, they enroll into a Common Core program that extends for two and a half years, after which they have to choose a major. However, the accreditation process formally involved each of the engineering programs offered, not the Common Core, which could have meant preparing more than a dozen different evaluations of the Common Core, as part of each of the individual programs. Fortunately, this could be avoided by factoring out this part of the collection of data. For the evaluation of the Common Core Program, the surveys were directed to students who had already finished that stage and to professors of the engineering programs who received the students coming out of the Common Core.

## Accreditation Visits

During the 3 days of accreditation visits, the peer-review committees evaluated three main aspects of each program: desired professional profile for the graduates and learning outcomes, operational conditions in place to achieve the desired purposes, and capacity for self-regulation needed to identify strengths and weaknesses. This evaluation was performed through general meetings with the authorities of the Faculty and School of Engineering, Faculty Council and School Council, Teaching

Development Committee, and students and professors of the Common Core program. After these initial meetings, during the first day the peer-review committee for each department held meetings with the self-study committee and employers. The second day was dedicated to meeting with the department's leaders (Chair, Department Council, and Teaching Council) and professors (full and part time), plus students of different levels and their leaders. Meetings with alumni were also held on the second day. The visiting committees also became familiar with the teaching and lab infrastructure available. Finally, the third day was dedicated to internal meetings of the committees and late that day the preliminary reports were communicated.

## *Accreditation Reports*

The accreditation reports, which came in April, confirmed the preliminary reports issued at the time of the visit. Most of the programs were accredited for the maximum allowed time (7 years), with a few exceptions that received 6 years of accreditation. The quality of the students and of the faculty was considered to be very good, as was the infrastructure (with few exceptions). There was general agreement about the positive contribution of the Common Core to the curricula of all the programs, as well as agreement on the need to develop more communication and teamwork skills for all graduates. It was also considered positively that the new curriculum had already identified that as a weakness, and included new courses and teaching methodologies aimed at addressing those needs.

In Chile there are 70 civil engineering programs offered by 18 universities. Only 9 of these programs have been accredited for the maximum 7 years, and 13 of them for 6 years. Of these, 5 out of the 9 programs accredited for 7-years and 4 out of the 16 for 6-years are from the University of Chile, which makes the School of Engineering and Science one of the best performers in the accreditation process.

## *Lessons Learned*

Going through this process was a learning experience for everyone involved in the university. Tasks such as collecting statistical data turned out to be much harder than expected, and pointed out the need to develop appropriate systems, not just for future accreditation processes, but for the school's own institutional analysis needs. Contacting relevant alumni and employers was also nontrivial, again pointing at weaknesses in the management of relations with these two important groups of stakeholders. Finally, having to gather explicit evidence and collect numerical data that would prove things that are used to be taken for granted (such as the professional success of the graduates) illustrated the need to have a clear statement of the school's intended goals and the need to use data to decide what changes should be made.

## Regional Overview

As for other Latin-American countries, while the development of Quality Assurance practices has followed different pathways, it is safe to say that it has become an equally important and challenging issue for the different higher education systems across the region. A brief review of the higher education Quality Assurance systems in Argentina, Brazil, Colombia, and Mexico provides a clear indication of this situation (for further information see CINDA 2007).

In the case of Argentina, Quality Assurance practices have been carried out by the National Commission of University Accreditation and Evaluation – CONEAU. It was established in 1995 as part of the Higher Education Law 24.521. This public organization is in charge of assessing the creation of new public and private institutions, state-regulated programs, graduate programs, as well as the recognition of private agencies in charge of university accreditation and assessment.

As for Brazil, accreditation is under the responsibility of federal and state government. The current federal project is the National System of Higher Education Evaluation – SINAES, which has three components: institutional, undergraduate programs, and student assessment (National Exam of Student Achievement – ENADE). The accreditation of graduate programs is the responsibility of the Coordination for Higher Education Personnel Training – CAPES, which has existed since 1976. The Colombian situation is characterized by the existence of various organizations that share different Quality Assurance functions. The Higher Education Quality Assurance System – SACES and the National Council of Accreditation – CNA are responsible for the voluntary accreditation of programs and institutions, whereas the National Intersector Commission for the Quality Assurance of Higher Education – CONACES, created in 2003, assesses and registers programs and institutions that fulfill required quality standards. The Colombian Institute for the Promotion of Higher Education – ICFES administers the State Examination of Higher Education Quality taken by upper-level undergraduate students.

As for Mexico, the Interinstitutional Committees for the Assessment of Higher Education – CIEES conduct voluntary external assessments to undergraduate and graduate programs. The Council of Higher Education Accreditation – COPAES is in charge of the official recognition and accreditation of programs, which is carried out by authorized private agencies. The National Centre for the Assessment of Education – CENEVAL administers examinations to senior undergraduate students.

## Conclusions

The present Quality Assurance system of Higher Education in Chile is the outcome of a long-term national learning process. Since its start, technical and political variables have been interwoven. Thanks to a balanced mix of expectations and flexibility, the system has progressed steadily, gaining acceptance from all political quarters.

The official agenda calls for further development of the Quality Assurance system in the areas of professional licensing, installation of a national system of higher education indicators, and, in general, increasing effectiveness of the Quality Assurance cycle to stimulate continuous improvement.

## References

Centro Interuniversitario de Desarrollo-CINDA. (2007). Educación Superior en Iberoamérica Informe 2007. Santiago, Chile: RIL Editores.
Comisión Nacional de Acreditación-CNAP. (2007). CNAP 1999–2007 El modelo chileno de acreditación de la educación superior. Santiago, Chile: CNAP, Ministerio de Educación.
Lemaitre, M. J. (2002). Quality as politics. Quality in Higher Education, 8(1), 29–37.
Letelier, M. F. (1997). Innovations in engineering education in Chile. 1997 ABET Annual Meeting Proceedings, Washington, DC.
Letelier, M. F., & Carrasco, R. (2004). Higher education assessment and accreditation in Chile: State-of-the-art and trends. European Journal of Engineering Education, 29(1), 119–124.
Letelier, M. F., Herrera, J. A., Canales, A. M., Carrasco, R., & López, L. L. (2003). Competencies evaluation in engineering programmes. European Journal of Engineering Education, 28(3), 275–286.
Letelier, M. F., & Sandoval, M. J. (2007). Acreditación e Ingeniería: Algunas áreas oficiales de innovación. Actas 22º Congreso Chileno en Educación en Ingeniería. Santiago, Chile: Universidad de Chile.

# Quality Assurance of Engineering Education in Sweden

**Johan Malmqvist and Aija Sadurskis**

**Abstract** The chapter describes the past and future for Quality Assurance programs in Sweden. The Swedish National Agency for Higher Education is the government agency responsible for the evaluation of the quality of university education in Sweden. One of the purposes of the most recent national evaluation of Swedish engineering degree programs in 2005 was to compare the programs and find examples of good practice. In the evaluation, the conceive–design–implement–operate (CDIO) self-evaluation model was introduced to the universities as a model for engineering education development and as an instrument for continuous self-improvement. The next generation of Quality Assurance in Sweden, to be applied in the 2011, is described, focusing on new and changed aims of the Quality Assurance program.

## Introduction

The Swedish National Agency for Higher Education (*Högskoleverket*) is responsible for the evaluation of the quality of university education in Sweden. In this chapter, we describe the agency's Quality Assurance program in some detail, as well as how it was applied in the 2005 national evaluation of Swedish *civilingenjör* engineering degree programs, when the conceive–design–implement–operate (CDIO) self-evaluation model was introduced to the universities. (For a discussion on CDIO, see also chapter "CDIO and Quality Assurance: Using the Standards for Continuous Program Improvement" by Brodeur and Crawley or refer to Crawley et al. 2007.) Recommendations and lessons learned from the use of the CDIO model, and the evaluation itself are also summarized. Finally, the chapter concludes with a description of the next generation of Quality Assurance in Sweden, focusing on new and changed aims and components of the Quality Assurance program.

J. Malmqvist (✉)
Department of Product and Production Development, Hörsalsvägen 7A,
Chalmers University of Technology, SE-412 96 Göteborg, Sweden
e-mail: johan.malmqvist@chalmers.se

A.S. Patil and P.J. Gray (eds.), *Engineering Education Quality Assurance: A Global Perspective*,
DOI 10.1007/978-1-4419-0555-0_10, © Springer Science+Business Media, LLC 2009

## Quality Assurance in Sweden

Sweden has 39 higher education institutions (HEI), some 300,000 undergraduate and master-level students, and some 18,000 doctoral students. In Sweden, as in many other developed countries, the higher education sector has expanded in the past decades. An increasing part of the population enrolls in higher education, approaching a goal of 50% of high school graduates enrolling. Higher education in Sweden can thus said to be moving toward a system of mass or majority education.

In 1993, Swedish HEI were given considerably increased powers and responsibilities for a number of issues. With this university reform, government by rules was replaced by government by objectives. However, with decentralization of responsibility and power comes greater accountability in that the higher education sector has to show how goals are met. Not surprisingly, demands for accountability, follow-up, and evaluation of higher education have grown. Several groups have begun demanding evidence of quality: politicians, tax payers, and students burdened with study loans.

In January 2001, a national program for Quality Assurance was introduced, with the Swedish National Agency for Higher Education (HSV 2008) evaluating subjects and programs.[1] All studies that lead to general or professional degrees were to be evaluated in a 6-year period, being the first of recurring cycles of evaluation.

The evaluations have two main aims: control and development. They can also serve other purposes, for example, to inform students or to lend authority. The evaluations are intended to contribute to the universities' internal quality and development work; the control component can be described as making certain that studies meet minimum requirements. There has, at any rate so far, been no attempt to rank the subjects or programs, since it has been believed that what is evaluated is too complex for ranking to be meaningful. There is also a connection between the evaluations and the right to award degrees. If serious quality flaws are noted, the university or university college should be aware that the right to award a degree can be revoked if no action is taken within a year.

## The Evaluation Model

The Swedish evaluation of higher education has a so-called theory-oriented approach. Theory-oriented evaluation includes the components' conditions, process, and results. The evaluation must elucidate and critically analyze all the three. By relating results to the preceding process and the pre-existing conditions, the evaluator can help explain why things are in the way they are.

The Swedish National Agency for Higher Education evaluation model has followed the internationally accepted pattern of self-evaluation, peer review by an external panel

[1] The Swedish National Agency for Higher Education (HSV) is a central agency, under the Ministry of Education, responsible for various matters relating to higher education. In addition to Quality Assurance, the Agency's main activities are supervision, analysis and information.

making a site visit, and a report. However, there are certain rather unique characteristics of the Swedish evaluation program: the importance of follow-up of evaluation is stressed, and students are given the role not only of informant, but also of expert.

The first step of the evaluations is been a self-assessment, made by the HEI that offers a particular degree. The self-assessment provides a background to the subject or program to be evaluated. *Self-assessment* implies that the HEI analyzes the studies offered, and identifies strengths and weaknesses. The self-assessment is carried out according to a format, stated by the Agency and including a number of topics to be covered. Some of these are teacher competence, educational goals, contents and organization of instruction and examination, as well as quality and availability of library and other sources of information. There is a common base for all self-assessments across the university sector, complemented by domain-specific issues.

The second element of the evaluation is a site visit by an external review panel. The self-assessment report is studied by the panel and discussed with the program management, faculty, other university-level staff, and students at a site visit. The purpose of the visit is to confirm and deepen observations made from studying the self-assessment reports. Based on the information from the self-assessments and interviews, the panel writes its report, describing strengths and weaknesses and making its recommendations.

The external panel is made up of experts from the field being evaluated, students, and, in some panels, representatives of industry or other relevant job market sectors. The latter is especially important in evaluations of study programs leading to professional qualification degrees, such as engineering degrees. The choice to include students in the panels, thus viewing them as stakeholders in addition to informants, was controversial at the outset of these evaluations, but is now actively encouraged by the European Association for Quality Assurance in Higher Education. All members of the panel are nominated by the institutions being evaluated.

The third element of the evaluation model is follow-up. A few months after the report is published, representatives of the evaluated subject or program are convened to discuss the contents of the report as well as the evaluation process. In addition, the results of the evaluation are followed up 1–3 years later with the purpose of assessing the effects of the evaluation's recommendations. The Swedish National Agency for Higher Education publishes annual reports analyzing the quality of education based on the evaluations that took place during the preceding year.

## Evaluation of Engineering Education

In Sweden, engineering education mainly takes the form of integrated programs leading to professional degrees. There are two such degrees: the *civilingenjör* degree, roughly corresponding to a Master of Science or *Diplom-Ingenieur* degree, is achieved after 5 years of study,[2] and the *högskoleingenjör* degree, a university diploma in engineering, is reached after 3 years. The latter is academically less demanding.

[2] 4.5 years at the time of the 2005 evaluation.

The *civilingenjör* programs are offered at 11 universities or university colleges. However, the number of programs at any one institution varies from 2 to 16. In total, approximately 100 *civilingenjör* programs are offered. They cover all areas of engineering science including information technology, engineering physics, chemistry, biotechnology, mechanical engineering as well as surveying. The 2005 evaluation described in this article was restricted to evaluation of *civilingenjör* engineering education.[3]

The 2005 evaluation was marked by a greater than usual involvement of stakeholders. The institutions to be evaluated formed a joint group well ahead of the start of the evaluation and this group had ongoing discussions with the agency. Stakeholders were also extensively involved, meeting the agency once before starting the evaluation, and once after the first two meetings of the external panel. These meetings served to define focal points and to discuss differences of opinion. As part of the agency's general policy to have a student perspective, meetings with groups of students were arranged on several occasions to hear their wishes and to discuss their involvement in the evaluation process. Also, of the eight people in the panel, two were students. An external stakeholder perspective was assured by including two representatives of industry in the panel, one of whom was the chairman.

Although the evaluation followed the general format of the agency's evaluations, there were a number of modifications. There were several reasons for this. One was that the *civilingenjör* programs had been quite extensively evaluated before, at their own initiative. Also, there was no reason to believe that the programs did not meet (at least) minimum requirements. It was therefore decided that this evaluation would have more of a benchmarking character. The attempt was to compare the programs and find examples of good practice. In order to do this, the evaluation procedure needed to be more standardized, using a more detailed and concrete self-evaluation manual than was normally the case. The manual was therefore changed to comprise 21 questions to be answered by the HEI centrally, and an additional 46 questions to be answered by each program.

## Overall Quality of *Civilingenjör* Engineering Education in 2005

Generally, the quality of the engineering education being evaluated was found to be good, resulting in engineers that are internationally competitive. Training in engineering and natural sciences was generally thought sufficient. However, socially, economically, and environmentally sustainable applications of technology were less well provided for. There were also shortcomings in the administration and management of the educational programs, and responsibility did not always match authority. While many excellent examples of dialog with future employers were found, there were also cases where more work practice should be included.

[3] The *högskoleingenjör* programs were evaluated in 2002.

Faculty qualifications were strong, especially research qualifications, but faculty members were hard pressed. They were forced to devote time that should have been used for research, as well as their own free time, to teaching.

Engineering education had expanded dramatically, with a large increase in the number of programs offered. At the same time, the number of students seeking admission had decreased, despite major recruitment efforts. The percentage of women enrolling, already comparatively low, had sunk further. The students took a long time to graduate and the HEI lacked sufficient systems for monitoring and increasing throughput. A general finding in the 2005 evaluation was that several learning environments were small and vulnerable. The incentives to obtain a realistic educational volume were found to be insufficient. Also, the type of programs offered was not always optimal in terms of employability. The evaluation panel concluded that there should be incentives for the HEI to specialize, to invest in the types of education that the labor market needs, and to get the students to complete their studies within a normal period of time.

## Inclusion of a CDIO Component

As noted above, the evaluation model used in 2005 followed the internationally accepted pattern of self-assessment, peer review by an external panel making a site visit, and a report. In addition, as noted, the 2005 evaluation had more of a benchmarking character than previous evaluations. Furthermore, the CDIO self-evaluation model was introduced in this evaluation, as a model for engineering education development and as an instrument for continuous self-improvement.

The basic structure of the self-assessment questions posed to the HEIs in 2005 is indicated in Fig. 1. The questions are divided into university-level questions and program-level questions, and then further decomposed into questions related to conditions, results, and processes. There are about 20 university-level questions and about 50 program-level questions (HSV 2004). One example of a university-level question is *How does the university use knowledge about and experiences from graduated students in its educational planning?* An example of a program-level question is: *Describe the program in terms of specific goals and profile(s). Account for the considerations made when designing the program. Attach the program plan.*

For this evaluation, the Swedish National Agency for Higher Education also decided to add an overall program assessment component to the questions (HSV 2004). The purposes were to:

- Complement the responses to the basic questions in order to attain a more comprehensive, overall assessment of the university and program
- Give the external review panel an additional instrument for its analysis and evaluation
- Provide the universities/programs with an instrument that can be applied as a basis for future continuous improvement efforts

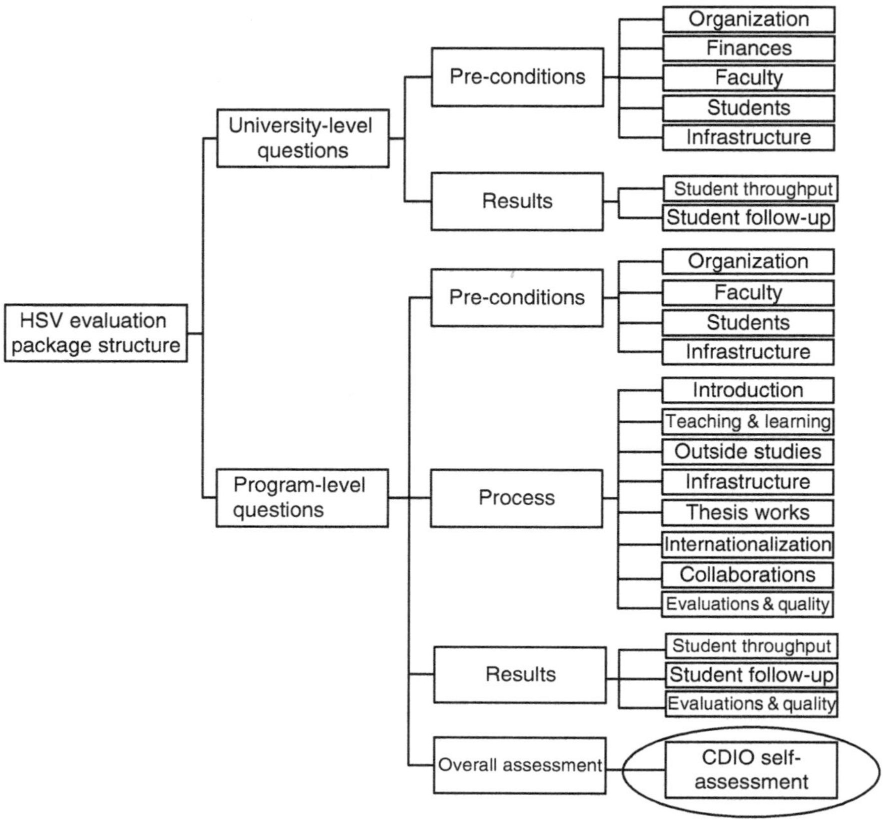


**Fig. 1** Structure of HSV self-assessment questions package

## CDIO Standards

The CDIO model (Berggren et al. 2003; Crawley et al. 2007) is a model for engineering education that stresses that the product lifecycle – CDIO – should form the framework for the design of the engineering educational program. The educational design process is guided by the CDIO standards, a set of 12 principles that characterizes this educational model as well as general good practice in education (Brodeur and Crawley 2005).[4]

The CDIO standards define the essential characteristics of an engineering program that has adopted the CDIO model of engineering education reform (Brodeur and Crawley 2005). (An asterisk denotes the seven *essential* standards.)

[4] See the Brodeur and Crawley chapter "CDIO and Quality Assurance: Using the Standards for Continuous Program Improvement" for a description of the CDIO project.

Standard 1: CDIO as context*
Standard 2: CDIO syllabus outcomes*
Standard 3: Integrated curriculum*
Standard 4: Introduction to engineering
Standard 5: Design–implement experiences
Standard 6: CDIO workspaces
Standard 7: Integrated learning experiences*
Standard 8: Active learning
Standard 9: Enhancement of faculty CDIO skills*
Standard 10: Enhancement of faculty teaching skills
Standard 11: CDIO skills assessment*
Standard 12: CDIO program evaluation*

The 12 standards were developed in response to the request from program stakeholders to be able to recognize CDIO programs and their graduates. The 12 CDIO standards serve as guidelines for educational program reform and evaluation, create benchmarks and goals with worldwide application, and provide a framework for continuous improvement. The 12 CDIO standards address program philosophy, curriculum development, design–build experiences and workspaces, new methods of teaching and learning, faculty development, and assessment and evaluation. Seven are considered essential because they distinguish CDIO programs from other educational reform initiatives; five supplementary standards significantly enrich a CDIO program and reflect best practice in engineering education.

The CDIO standards and the associated self-assessment tools (Brodeur and Crawley 2005) were chosen for the purposes noted above. However, a number of modifications were also made to adapt the standards to the Swedish higher education and engineering education context. First, the standards were reformulated to avoid the use of the acronym *CDIO* while keeping the corresponding content. Second, the programs were also given an option to restate Standard 1, enabling them to replace the *product and system development* context with another more fitting to their particular program. Finally, there was no summary of a total score, the intention being to keep HEI from using their total score as a basis for some kind of ranking.

The fulfillment of each standard is measured by a five-level scale, thus also providing a tool for continuous improvement. So far, the CDIO standards have been applied to a limited number and range of educational programs, essentially the collaborators in the CDIO Initiative (2008). This was the first large-scale application of the CDIO standards in a national evaluation.

The determination of a program's progress toward fulfillment of the CDIO standards is accomplished through self-evaluation. An excerpt of the layout of the self-evaluation form is shown in Fig. 2. The fulfillment of each standard is measured by a five-level scale, which is used to rate the progress toward the planning, implementation, and adoption of each CDIO standard.

The rubrics of the five-level scale are stated in Fig. 3.

Self-assessment using the 12 CDIO standards and the five-level rating scale provides a tool for the monitoring of improvements via a series of evaluations where overall program improvement can be made visible through an increase in

**Compliance with CDIO Standards**

**Institution:**
**Program:**
**Evaluators:**
**Date:**

| | CDIO STANDARD | EVIDENCE OF COMPLIANCE | RATING | ACTIONS |
|---|---|---|---|---|
| 1 | Adoption of a mission statement that includes the principle that product and system life cycle development and deployment–Conceiving, Designing, Implementing and Operating - are the context of engineering education* | | | |
| 2 | Specific, detailed goals for personal, interpersonal and product and system building skills, consistent with program mission and validated by program stakeholders* | | | |
| 3 | A curriculum designed with mutually supporting disciplinary subjects, with an explicit plan to integrate personal, interpersonal and product and system building skills* | | | |
| 4 | An introductory course that provides the framework for engineering practice in product and system building, and introduces essential... personal and interpersonal skills | | | |

**Fig. 2** Excerpt from conceive–design–implement–operate (CDIO) self-evaluation form

| | |
|---|---|
| 0 | No initial program-level plan or pilot implementation |
| 1 | Initial program-level plan and pilot implementation at the course or program levels |
| 2 | Well-developed program-level plan and prototype implementation at the course or program levels |
| 3 | Complete and adopted program-level implementation of the plan at the course or program level under way |
| 4 | Complete and adopted program-level and comprehensive implementation of the plan at the course or program levels, with continuous improvement processes in place |

**Fig. 3** Rating scale used in self-evaluation with the CDIO standards

total score. In order to facilitate application of the CDIO standards self-assessment procedure, the programs were supplied with a set of instructional documents (translated into Swedish), including:

- The description of the CDIO standards
- A set of headings and topics for a program goal statement, essentially a condensed version of the CDIO syllabus (Crawley 2002)
- A template for the evaluation form
- Two examples of CDIO self-assessments

## Experience Gained from the Use of CDIO Standards

As has been described above, in the Swedish national evaluation of engineering degree programs a modified version of the CDIO standards was used to evaluate about 100 engineering programs.

From the evaluator's perspective, the use of the CDIO element brought in a selection of topics guided by the CDIO standards that helped focus on key issues while the associated quantitative rating scale standardized the response format. This was essential to enable some systematic comparisons across the large number of programs that took part in the evaluation. Examples of general recommendations in the evaluation report that were grounded in such data include requests for more distinct program goals with explicit connections to course goals, for introductory courses, and for pronounced strategies for forms of instruction and examination, linking them to the different types of knowledge and skills of the program (HSV 2006).

Recommendations based on the CDIO self-assessment data were also part of the program-specific conclusions that were communicated to each program, giving these recommendations a structured and consistent format. There was also a strong connection between the recommendations and characteristics of programs and universities that were identified as *good* examples, including aspects such as inclusion of CDIO elements, well-developed programs for faculty competence development, and contacts between students and future employees. This clarified the basis for the recommendations as the review panel articulated a view of *good* engineering education, increasing the transparency of the analysis and recommendations.

As a separate initiative, a survey and interview study was carried out to investigate the program directors' view of the relevance, limitations, and ease of use of the CDIO standards. The survey questionnaire was divided into five parts. The first part covered background questions concerning what type of program the respondent represented and previous knowledge of the CDIO Initiative. In the second and third part, the respondents were asked to judge the ease of understanding, the ease of use, the relevance and the applicability of the overall CDIO standards as well as each individual CDIO Standard. The fourth part of the questionnaire covered the rating scale, and finally the respondents were given the opportunity to give general comments on positive and learning aspects of the CDIO standards and also to suggest improvements to the standards. The quantitative data were complemented with qualitative data obtained from interviews with selected program managers, chosen to represent program types that had not earlier been involved in the CDIO Initiative.

The results of the survey are reported in detail in Malmqvist et al. (2006). Survey and interview results indicate that the standards are relevant and applicable for a wider range of programs than had earlier used them, and that changing toward implementing the standards would improve program quality. The survey results also indicate that the standards' most important benefit is that they provide a basis for systematic program development.

Challenging issues when doing a CDIO standards-based self-evaluation revealed by the survey include interpreting Standard 1 in the context of the science and technological domain in question and the proper use of the rating scale. There are also concerns that the fact that the program's actions to develop personal, interpersonal, and product and system development and deployment skills are most evident in the evaluation does not do justice to its attention to disciplinary knowledge and connections to research. These results suggest a need to complement the CDIO standards with other instruments in an overall evaluation, and to make its role in the context clear.

## The Next Generation of Quality Assurance in Sweden

The Swedish institutions of higher education and the Swedish National Agency for Higher Education now find themselves at the beginning of a second 6-year cycle of evaluations. As with the first cycle, various stakeholders have been consulted when planning the subject and program evaluations. It has been a tenet of these discussions that the design of the second cycle should not be identical to the first, since simply repeating an evaluation will not yield an equally good result. Also, the institutions of higher education may be assumed to shoulder a greater part of the responsibility for educational quality. Certainly, their own Quality Assurance systems have benefited considerably from the experience of the past 6 years.

The design of the second cycle finally decided on is that evaluations will be conducted in three stages. Initially, a national picture of subjects or programs will be drawn up. This picture will contain information from simplified self-assessments as well as standardized key data. Based on this picture, possibly supplemented with information from other sources, a selection will be made of subjects or programs to undergo further scrutiny, the second stage. The choice may be of all institutions of higher education, or, depending on the types of problem anticipated, of scrutinizing some institutions rather than others. The choice will be made by the agency with the aid of subject experts. In the third stage, in-depth evaluation will be carried out according to the same model that has been used so far, including a site visit by a panel of external assessors. The evaluation relies on predetermined aspects and criteria, with the assessors summarizing their findings in a report. This report would form the basis for decisions on any possible sanctions by the agency. The *civilingenjör* engineering programs are scheduled to be evaluated again in 2011. Whether they will be evaluated according to the same format, including the CDIO component has not yet been decided.

A Government White Paper, published in November 2007, on the allocation of resources for research and higher education, has suggested that a portion of Government funding should be linked to quality indicators. It is not yet clear what the Government will do with this suggestion, but a link between quality appraisal and resource allocation poses a new challenge to the evaluations of educational quality with the distinct possibly of ramifications for individual HEI offering engineering education.

## References


Berggren, K. F., Brodeur, D. R., Crawley, E. F., Ingemarsson, I., Litant, W. T., Malmqvist, J., et al. (2003). CDIO: An international initiative for reforming Engineering Education. *World Transactions on Engineering and Technology Education, 2*(1), 49–52.

Brodeur, D. R., & Crawley, E. F. (2005). Program evaluation aligned with the CDIO standards. *Proceedings of ASEE-05, Portland, Oregon.*

CDIO Initiative. (2008). CDIO initiative Web site. Accessed on October 1, 2008. http://www.cdio.org

Crawley, E. F. (2002). *The MIT CDIO syllabus report: Technical report.* The CDIO initiative Web site. Accessed on October 1, 2008. http://www.cdio.org

Crawley, E. F., Malmqvist, J., Östlund, S., & Brodeur, D. R. (2007). *Rethinking engineering education: The CDIO approach.* New York, NY: Springer.

HSV. (2004). *Utvärdering av civilingenjörsutbildning. Anvisningar och underlag för självvärdering (Evaluation of engineering programs. Instructions and forms for self-assessment).* Stockholm: HSV.

HSV. (2006). *Evaluation of civil engineering programs at Swedish universities and institutions of higher education.* Report 2006:31 R. Stockholm: HSV.

HSV. (2008). Swedish national agency of higher education Web site. Accessed on October 1, 2008. http://www.hsv.se

Malmqvist, J., Edström, K., Gunnarsson, S., & Östlund, S. (2006). The application of CDIO standards in the evaluation of Swedish engineering degree programmes. *World Transactions of Engineering and Technology, 5*(2), 361–364.

# Assessment of Engineering Education Quality: An Indian Perspective

**R. Natarajan**

**Abstract** This chapter explores the differences in perception of Quality in the Manufacturing, Services, and Education sectors in India. The anatomy of an Engineering Institution is then examined. Some characteristics of institutions of excellence are highlighted and the frameworks for Quality Assurance currently used in India are analyzed. This is set within the context of the criteria and weights employed by several assessments of ranking of Academic Quality within India.

## Introduction

What does academic quality mean? Most experts agree that the top-class institutions invariably possess several common characteristics. Academic quality can be described in terms of the educational environment, mission, and clarity of purpose of an institution. Its description can also include input variables such as high-quality faculty; excellent physical facilities, in the form of classrooms and laboratories; adequate resources to maintain the operation; curricula with a variety and depth of courses with appropriate rigor; and adequate number and mix of students to enable students to learn from one another and maintain individualized learning. Most importantly today is the description of academic quality related to outputs, i.e., the learning of its students while in college and their success upon graduation. Research output by faculty members may also be an indicator of quality in those institutions with research as a mission. There are reciprocal relationships among the quality of educational environment, the input variables, and the output variables, with each element enhancing and reinforcing the others.

R. Natarajan (✉)
52/1, 13th Cross Road, Malleswaram, Bangalore 560 003, India
e-mail: prof.rnatarajan@gmail.com

A.S. Patil and P.J. Gray (eds.), *Engineering Education Quality Assurance: A Global Perspective*,
DOI 10.1007/978-1-4419-0555-0_11, © Springer Science+Business Media, LLC 2009

## The Evolving Concept of Quality Assurance

Historically the concept of Quality Assurance or quality control has been the focus mainly of the manufacturing sector (Prahalad and Krishnan 1999). In those industries, quality was about minimizing defects and ensuring that the manufactured products conformed to clear specifications.

Service businesses have had to develop a more comprehensive view of quality (Prahalad and Krishnan 1999). They are not only concerned with minimizing defects (i.e., errors or bad service), but also concerned with managing expectations, experiences, and emotions since these may differ for each consumer.

## Implications for Engineering Education

The US National Science Foundation (NSF) Task Force on Total Quality Management (TQM) (Grant 1993) has the following definition of Quality Engineering Education:

- Quality Engineering Education is the development of intellectual skills and knowledge that will equip graduates to contribute to society through productive and satisfying engineering careers as innovators, decision-makers, and leaders in the global economy of the twenty-first century.
- Quality Engineering Education demands a process of continuous improvement of and dramatic innovation in student, employer, and societal satisfaction by systematically and collectively evaluating and refining the system, practices, and culture of Engineering Education Institutions

The Task Force pointed out that TQM is not a destination, but rather a journey to improvement. The Task Force has also examined the nature of the customers of Engineering Education or stakeholders, which can vary widely depending on an Institution's mission, goals, strategies, and tactics. The stakeholders include suppliers, such as high schools, the students themselves, and their parents, and receivers, such as employers and society in general.

## Anatomy of a Research University

Figure 1 shows the processes which constitute the core activities of a research university [such as the Indian Institutes of Technologies (IITs)], the three major categories of inputs, and the two major classes of outputs. While the tangible outcomes are indeed necessary for every institution and are easy to measure, it is the intangible outcomes that differentiate the best from the mediocre.

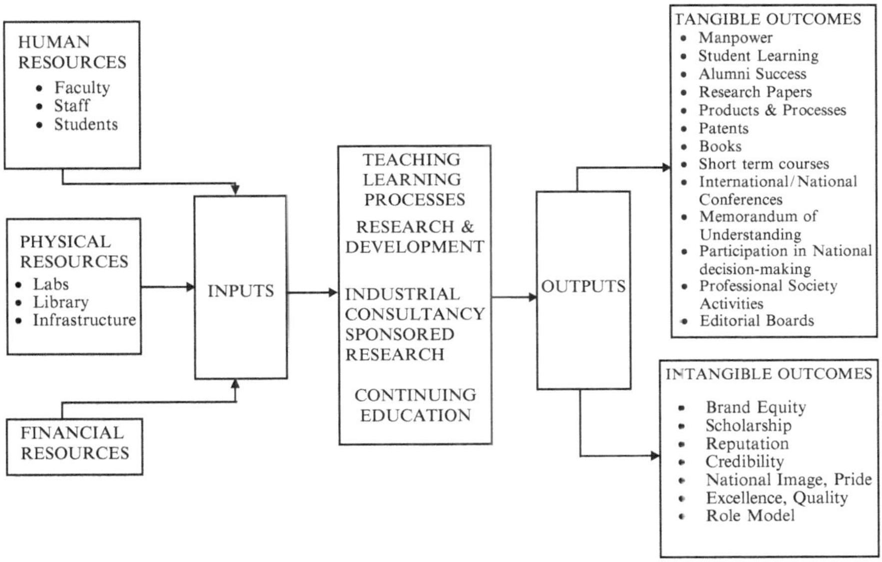


**Fig. 1** Anatomy of a research university

## Some Characteristics of Institutions of Excellence

Some studies (Rice and Austin 1988; Heverson 1987) have been made on *what exemplary colleges do right*. Some of the conclusions are discussed below. One of the key examples of what exemplary colleges do right is that the faculty of such institutions are deeply committed to their work, and support their institutions' distinctive missions enthusiastically.

Four key features were identified in these institutions as the sources of faculty morale and satisfaction:

- They all had distinctive organizational cultures that were carefully articulated, nurtured, and sustained.
- They had strong participatory leadership which provided direction and purpose.
- They had a firm sense of organizational momentum; they were institutions on the move.
- The faculty of these institutions had an unusually compelling identification with the institution.

## Frameworks of Quality Assessment

Two frameworks for quality assessment of technical/higher education in India are briefly discussed in this section.

## *The Accreditation Initiatives of the National Board of Accreditation*

The first framework is the initiative of the National Board of Accreditation (NBA). The NBA was established in September 1994, following the conferment of statutory status to the All India Council for Technical Education (AICTE) in 1988. The NBA has a mandate to periodically conduct evaluation of technical institutions or programs on the basis of guidelines, norms, and standards specified by it and to make recommendations to it, or to the Council, or to the Commission or to other bodies, regarding recognition or derecognition of the institution or the program (All India Council for Technical Education Act 1988). The NBA has been charged with the task of evolving a procedure for quality assessment in the technical education sector and specifically to:

- Articulate the criteria for assessment of quality
- Identify parameters to quantitatively assess these criteria and assign appropriate program-specific weights for each
- Validate the procedure by well-designed test runs
- Establish appropriate benchmarks

In order to carry out its charge, the NBA has provided awareness workshops, training programs, and other essential activities, such as benchmarking and finalization of the evaluation procedures and methodologies, throughout the country.

The major policy decision adopted by the NBA is to accord Accreditation, not to the institutions as a whole, but at the program level, i.e., the 3-year Diploma program, the 4-year under-graduate engineering degree program, and the 4-semester Master of Engineering/Master of Technology (ME/MTech) program. The final decision on accreditation is based on (1) the self-assessment provided by the Institution, (2) a 3-day visit by peer assessors and their subsequent report, (3) evaluation of the report by a Sectorial Committee, and (4) recommendations by the Committee to the National Board of Accreditation.

Initially, the programs were graded A, B, C, and NA (not accredited), depending on their rating on a 1,000-point scale. Subsequently, to maintain parity with Washington Accord countries, this policy has been given up, and a *yes* or *no* decision is accorded where a distinction is made between programs securing greater than 750 points, which are given accreditation for 5 years; those securing between 650 and 750 points, which are given accreditation for 3 years; and those securing less than 650 points which are not accredited. Table 1 shows the NBA Accreditation criteria, divided into eight categories, amounting to a total of 1,000 points.

It can be seen that the points are different for Diploma, Under-Graduate (UG), and Post-Graduate (PG) programs. The weight for the teaching–learning processes is highest for Diploma and least for PG programs, while the weight for research and development and interaction is highest for PG programs and least for Diploma programs.

**Table 1** National Board of Accreditation (NBA) accreditation parameters and weights

| | | Marks | | |
|---|---|---|---|---|
| | Parameter | Diploma | UG | PG |
| I | Organization and governance | 30 | 80 | 50 |
| II | Financial resources: Allocation and utilization | 70 | 70 | 50 |
| III | Physical resources (central facilities) | 50 | 50 | 50 |
| IV | Human resources – faculty and staff | 200 | 200 | 200 |
| V | Human resources – students | 100 | 100 | 100 |
| VI | Teaching–learning processes | 450 | 350 | 250 |
| VII | Supplementary processes | 50 | 50 | 50 |
| VIII | R&D interaction effort | 50 | 100 | 250 |
| Total | | 1,000 | 1,000 | 1,000 |

One of the main concerns of the Washington Accords related to accreditation is the explicit description of outcomes of the programs. Even though, in the current version, NBA accreditation does not address this aspect explicitly, the following quality indicators have been stated:

- *Institution.* Accreditation status and rankings (*institutional environment and institutional inputs*)
- *Faculty.* Performance in their multiple roles of teaching, research, and interaction with industry, the corporate sector, society, and the government. (*faculty inputs and research and teaching processes*)
- *Students and alumni.* Performance in university examinations and placement of graduating students in terms of both quantity (how many) and quality (where) (*outputs in terms of student learning and success*)

Approval or *recognition* by AICTE is necessary before a technical institution comes into existence. After two cohorts of students have graduated from the institution, it becomes eligible to be accredited by NBA. The major differences between these two processes are shown in Table 2.

## *Accreditation Initiatives of NAAC*

The National Assessment and Accreditation Council (NAAC) was set up as a Society by the University Grants Commission (UGC) in 1994. It covers *general nonprofessional* education, including universities and autonomous colleges. However, some engineering colleges have also sought accreditation from NAAC. Unlike NBA, which accredits programs, NAAC assesses and accredits the entire university or college.

Effective since April 2007, NAAC has a new methodology of assessment and accreditation, which has been designed with a view to overcome some of the

**Table 2** Some differences between recognition and accreditation processes

| Recognition | Accreditation |
|---|---|
| Performed before an institution is to be started | Performed after two batches have graduated |
| Fulfillment of initial conditions | Fulfillment of minimum norms of achievements and results |
| Assessment of promise | Assessment of performance |
| Largely based on physical, financial and infrastructural resources | Includes availability and quality of human resources, in addition |
| Based on project report | Based on information in self-assessment questionnaires which demand a clear articulation of Mission and Goals, and a SWOT analysis |
| Essentially a *quantity* assessment | Essentially a *quality* assessment |
| Reasonably straightforward | Much more complex |
| Decision : yes or no | Decision : grading into three classes |
| Not a new concept | A new concept in the Indian higher and professional education scene |

**Table 3** NAAC criteria and weights

| Criteria | University | Autonomous college | Affiliated/constituent college |
|---|---|---|---|
| I. Curricular aspects | 150(15%) | 100(10%) | 50(5%) |
| II. Teaching–learning and evaluation | 250(25%) | 350(35%) | 450(45%) |
| III. Research consultancy and extension | 200(20%) | 150(15%) | 100(10%) |
| IV. Infrastructure and learning resources | 100(10%) | 100(10%) | 100(10%) |
| V. Student support and progression | 100(10%) | 100(10%) | 100(10%) |
| VI. Governance and leadership | 150(15%) | 150(15%) | 150(15%) |
| VII. Innovative practices | 50(5%) | 50(5%) | 50(5%) |
| Total score | 1,000 | 1,000 | 1,000 |

limitations of its earlier methodology and to enhance its rigor, reliability, and validity. These changes are intended to significantly reduce interteam assessment variations, to make the process more user-friendly, and to enable NAAC to conduct the assessment of large numbers of institutions effectively and in a short time. Table 3 shows the NAAC assessment and accreditation criteria for universities, autonomous collages, and affiliated/constituent colleges.

The major differences between the weights for the three types of institutions are that areas of curriculum are more heavily weighted at the university level; teaching–learning and evaluation are less heavily weighted at the university level; and research, consultancy, and extension are more heavily weighted at the university level.

## Some Positive Consequences of Accreditation in India

There have been several beneficial consequences of NBA and NAAC accreditation initiatives. Every technical and higher educational institution has begun to appreciate the need to incorporate *quality* in its academic activities. The accreditation criteria provide guidelines to the institutions for achieving *quality* and *excellence*, and, in fact, define the profile of an *institution of excellence*. For example, since the criteria demand that every institution should have vision, mission, and goals; industry–institute interaction; research and development, etc., every institution strives to incorporate these into its portfolio of policies, plans, and activities. A healthy competition is evolving among the institutions, which share *best practices*, and seek to emulate the best institutions. The public, funding agencies, employees, and, in fact, all the stakeholders have started to appreciate the role of accreditation in promoting Quality Assurance in technical and higher education.

## References

All India Council for Technical Education Act. (1988).

Grant, H. (1993). The role of quality concepts in engineering education. Frontiers in Education Conference. *Proceedings Twenty-Third Annual Conference, Engineering Education: Renewing America's Technology, 6–9 November 1987* (pp. 535–539).

Heverson, E. D. (1987). Boosting academic reputation: A study of university departments. *The Review of Higher Education, 11*, 177–197.

Prahalad, C. K., & Krishnan, M. S. (1999). The new meaning of quality in the information age. *Harvard Business Review, 77*, 109–120.

Rice, R. E., & Austin, A. E. (1988). High faculty morale – What exemplary colleges do right. *Change, 20*(2), 51–58.

# Quality Issues Facing Malaysian Higher Learning Institutions: A Case Study of Universiti Teknologi Malaysia

**M. Puteh, S.M. Daud, N.H. Mahmood, and N.A. Azli**

**Abstract** This chapter analyses the extent to which Malaysian universities have responded to the pressing call for enhancing institutional quality and advocating academic excellence. Malaysian universities have been challenged by both internal and external assessments that suggest that they are not on a par with their international competitors. As a result there have been great demands for an urgent remedy. This chapter explores the progress that has been made in responding to those demands and the limitations that have been encountered. In so doing it provides a case study that focuses on engineering programmes in particular. The chapter is intended to answer the question: can institutional quality improvement deliver the human capital that Malaysia needs in order to remain globally competitive?

## Introduction

Higher education is a crucial ingredient in a country's success in the context of economic globalisation. Malaysia's goal of becoming a developed country by 2020 and, therefore, an important player in the global economy is very much dependant on its system of higher education. In recent years, Malaysian universities have been the centre of a public debate pertaining to their performance. For example, the standing of Malaysian universities in the *Times Higher Education Supplement* rankings has gradually deteriorated (Ariff 2007). The deteriorating standing of Malaysian universities in such a world-class ranking is a major cause of concern on the part of the Ministry of Higher Education (MOHE). MOHE has attributed this shortcoming to the poor quality of these institutions. In addition, public apprehension over the quality of Malaysian graduates is based on the fact that of the 10.275

M. Puteh (✉)
College of Science and Technology, Universiti Teknologi Malaysia, International Campus Jalan Semarak, 54100 Kuala Lumpur, Malaysia
e-mail: marlia@citycampus.utm.my

A.S. Patil and P.J. Gray (eds.), *Engineering Education Quality Assurance: A Global Perspective*,
DOI 10.1007/978-1-4419-0555-0_12, © Springer Science+Business Media, LLC 2009

million employed persons in Malaysia, only 1.9752 million have tertiary education. This is set within the context of a 3.3% unemployment rate in 2006 (Department of Statistics Malaysia 2006). Of course the question remains, is there a clear relationship between Malaysian graduate unemployment and the perceived poor quality of university programmes?

Never the less, Malaysian universities have been called on to respond to the perceived need to enhance institutional quality and to advance academic excellence. In order to address these goals, a Quality Assurance approach has been adopted in Malaysia and applied to all institutions and programmes, including engineering education. At the end of this chapter, a case study is presented that illustrates how the approach has been implemented in a Faculty of Electrical Engineering at a Malaysian public university.

## Quality Assurance in Malaysia

Extensive efforts to improve the quality of higher education have been taken in Malaysia. These involved systematic changes in policies, strategies, attitudes, procedures and activities by the Malaysian Qualifications Agency (MQA), previously known as the National Accreditation Board (LAN). The National Accreditation Board Act was passed on 26 September 1996 to ensure that high academic standards, quality and control are maintained in public and private higher education (Puteh 2006). Accreditation in Malaysian private colleges and universities was initially conducted by LAN. The Board conducted two types of evaluations: compliance of minimum standards and full accreditation of a course of study or programme. Private educational institutions are given a specific cycle (from 2 to 3 years for diploma and from 3 to 5 years for degree courses) to document their performance in these courses prior to the approval by LAN. Subsequently, the Quality Assurance Division (QAD) at the MOHE was formally established in April 2002 to manage and coordinate the Quality Assurance system for Malaysian public universities (Ministry of Higher Education 2005).

In 2002, the unemployment rate among private university graduates was 17% compared with 78% from public university graduates (Government of Malaysia 2002). Subsequently, a new entity known as Malaysian Qualification Agency (MQA) was created in November 2007 to merge the roles of the LAN and QAD. Its assignment is to supervise the Quality Assurance for both the public and private universities in Malaysia.

Accreditation functions as a kind of consumer protection whereby the public is assured of the general level of competence or achievement of students in a particular course of study within an accredited programme or institution (Prados et al. 2005). The Malaysian Quality Assurance scheme involves specialised accreditation whereby the accreditation is aimed at evaluating the details of programmes that prepare graduates for their professions, including engineering. Hence, the main role of the MQA is to implement the Malaysian Qualifications

Framework (MQF)[1] as the basis of Quality Assurance in higher education and as point of reference for the criteria and standards for national qualifications (Ministry of Higher Education 2005).

## Overview of Current Quality Assurance Scheme

Using the MQF, the MQA is responsible for charting new directions for Quality Assurance in higher education in order to provide the proper level of accreditation oversight for Malaysian universities. The following are the functions of the Malaysian Qualification Agency (2008):

1. To implement MQF as a reference point for Malaysian qualifications
2. To develop standards and credits and other relevant instruments as national references for the conferment of awards
3. To assure the quality of higher education institutions and their programmes
4. To accredit courses which fulfil the set criteria and standards
5. To facilitate the recognition and articulation of qualification
6. To maintain the Malaysian Qualifications Register (MQR)

Figure 1 illustrates the elements that are associated with MQA and how these components are consistent with the objectives of the MOHE.

In addition, a code of practice with criteria and standards for higher education in Malaysia was developed by MQA in June 2005. This code of practice is benchmarked against international *good practices* and provides guidelines and procedures for Quality Assurance in the following areas (Ministry of Higher Education 2005):

1. Vision, mission and learning outcomes
2. Curriculum design and delivery
3. Student selection and support services
4. Assessment of students
5. Academic staff
6. Educational resources
7. Programme monitoring and review
8. Leadership, governance and administration
9. Continual quality improvement

---

[1]The Malaysian Qualifications Framework (MQF) is Malaysia's declaration about its qualifications and quality in relation to its education system. MQF is an instrument that develops and classifies qualifications based on a set of criteria that are approved nationally and benchmarked against international best practices. It clarifies the earned academic levels, learning outcomes of study areas and credit system based on students' academic load. These criteria are accepted and used for all qualifications awarded by recognised higher education providers. Hence, MQF integrates with and links all national qualifications (Ministry of Higher Education 2005).

**Fig. 1** Contributions of MQA (Reproduced with the kind permission of the Malaysian Qualification Agency)

In general, there are two types of accreditations performed by the MQA:

*Provisional accreditation.* The initial process of accreditation which assists higher education providers to obtain full accreditation by enhancing the standard and quality established during the evaluation of provisional accreditation.
*Accreditation.* The full recognition which various higher educational institutions seek to achieve for their undergraduate and postgraduate degree programmes.

The MQA Act of 2007 also allows those higher institutions that have well-established internal Quality Assurance mechanisms to confer a self-accrediting status. To be so conferred, the higher education institution needs to undergo an institutional audit, and if successful, all qualifications it offers will be automatically registered in the Malaysian Qualification Registry (MQR). Besides assisting MQA in the registration of accredited programmes and qualifications, MQR is also a reference point for credit transfers of these programmes.

The functions and roles of MQA in enhancing the Quality Assurance process of Malaysian university programmes are rather broad as it covers all branches of learning. In addition, engineering programmes are subjected to another Quality Assurance organisation, the Malaysian Engineering Accreditation Council (EAC). The following section describes the functions of EAC and its requirement for accrediting engineering programmes.

## Engineering Education Quality Assurance

These recent Malaysian accreditation approaches have been adopted from those in the United States that have their roots in the historical development of engineering education over the last 70 years. By the early 1900s, engineering education advanced from two branches of learning: the formal mathematical-scientific system and the apprenticeship system. Most engineering schools in the United States combined both these approaches, resulting in an uncomfortable compromise in many institutions.

Formal accreditation of engineering programmes was first undertaken by the American Institute of Chemical Engineers (AIChE) in 1922 (Prados et al. 2005). Other countries followed including the Canadian Engineering Accreditation Board, the Engineering Council of the UK and the National Board of Accreditation, India (Patil and Pudlowski 2004).

The Malaysian EAC is designated by the Board of Engineers Malaysia (BEM) as responsible for accrediting engineering degrees offered by Malaysian universities. BEM registers graduates and professional engineers under the Registration of Engineers Act 1967 (revised 2002) and the pre-requisite for registration as a graduate engineer is any qualification in engineering recognised by the Board (Engineering Accreditation Council 2007). Therefore, the BEM's process of engineering programme accreditation is essential for ensuring that registered engineers achieve minimum competence comparable to global standards (Engineering Accreditation Council 2007).

In an effort to ensure that Malaysian engineering graduates are employable anywhere in the world, Malaysia, represented by EAC as its signatory organisation, applied to be a provisional member of the Washington Accord in 2001. Established in 1989, Washington Accord is an international agreement among bodies responsible for accrediting engineering degree programmes. It recognises the substantial equivalence of programmes accredited by those bodies and recommends that graduates of programmes accredited by any of the signatory bodies be recognised by the other bodies as having met the academic requirements for entry to the practice of engineering (International Engineering Alliance 2007).

As of 2003, Malaysia has already been accepted as a provisional member. Consistent with the objective of being accepted as a full signatory member of Washington Accord, the EAC has included in its accreditation manual criteria to ensure a continual quality improvement culture in the spirit of outcome-based education[2] (Engineering Accreditation Council 2007). The draft of a new Engineering Accreditation Council Manual had been created as of 2005 with a shift in emphasis on accreditation requirements toward outcome-based education. The

[2]Outcome-Based Education involves the shift in focus from curricula, resources and processes to outcomes and objectives. This places greater emphasis on the involvement of the stakeholders in establishing and measuring the outcomes and objectives of a particular engineering program. Outcome-Based Education focuses on students' learning by using learning outcome statements to explicitly state what the students are expected to know, understand or perform. This relates to the provision of learning activities which can ensure that the outcomes are achievable by the students. Most importantly, the extent to which the students meet these outcomes must be assessed through the use of explicit assessment criteria. Based on the assessment criteria, if the outcomes are not met, certain measures must be carried out for the purpose of continuous improvement (Mohammad 2007).

2005 Manual was drafted based on the recommendations of Malaysia's Washington Accord sponsors namely the Institution of Engineers, Australia (IEAust) and the Engineering Council, United Kingdom (ECUK) after a visit made in 2002. Further recommendations were made by IEAust, Accreditation Board for Engineering Technology (ABET, Inc.) and Hong Kong Institution of Engineers (HKIE) as Malaysia's Washington Accord mentors after their visit in 2004 (Wan Badaruzzaman 2005).

What then is the relationship between MQA and EAC and how do these two Quality Assurance bodies comply with the Quality Assurance system in Malaysia? The following section discusses the roles of EAC with regards to the undertaking of MQA in accrediting Malaysian university programmes, particularly the engineering curriculum.

## EAC Accreditation Requirements

EAC members are the representatives from the Board of Engineers Malaysia (BEM), the Institution of Engineers, Malaysia (IEM), Malaysian Qualification Agency (MQA) and the Public Services Department (PSD). Universities that propose to offer engineering programmes must submit a complete set of documents, through the MQA, for evaluation by the Malaysian EAC, as specified in the EAC Manual. That is, EAC assesses the documents and forwards a recommendation to the relevant authorities with the decision sent to the respective universities, through the MQA, with copies to the other Council members.

Accreditation is accorded to specific programmes by evaluating the nature of the programme, through the review of the set of EAC documents, and its institutional location, after the applicant university has satisfied all rules and regulations required. Accordingly, the EAC Engineering Programme Accreditation Manual of 2007, the assessment of an engineering programme involves two stages of evaluations: first, a review of the qualifying requirements of the university as outlined below and, second, a review of detailed assessment criteria related to the programme's academic curriculum, students, academic and support staff, facilities and quality management systems (Engineering Accreditation Council 2007). The EAC Components of Qualifying Requirements include:

1. Minimum 120 credit hours of which 80 credit hours must be core engineering courses offered over a period of 4 years
2. Final year project (minimum 6 credit hours)
3. Industrial training (minimum of 2 months)
4. Full-time academic staff (minimum of eight)
5. Staff to student ratio 1:25 (or less)
6. External examiner's report (minimum of two reports over 4 years)
7. Programme objectives
8. Programme outcomes

In addition, there are criteria related to evidence of the actual accomplishment of programme educational objectives and programme outcomes (Engineering Accreditation Council 2007).

The initial assessment of university qualifying requirements is meant to screen out programmes that do not meet the core requirements of the assessment criteria. Failure to meet any one of the eight qualifying requirements will disqualify the programme from further assessment by the EAC (Engineering Accreditation Council 2007). Once all the qualifying requirements are fulfilled, the programme is eligible to be evaluated, based on the accreditation criteria highlighted in the second stage of evaluation. The following section explores the experience of Universiti Teknologi Malaysia, Faculty of Electrical Engineering regarding the process of Quality Assurance.

## Case Study in Quality Assurance

The Universiti Teknologi Malaysia (UTM) began in 1906 as a class for technical studies at Kuala Lumpur City Council Building. Over the last 100 years the University has continued to grow and change until it finally became what it is today, as a university that focuses on producing graduates in the area of science and engineering. UTM has two campuses: the main campus located in Skudai, Johor and the branch campus situated at Jalan Semarak, Kuala Lumpur. The Skudai Campus offers degree as well as postgraduate programmes in a great variety of science and engineering areas while the branch campus, known as the City Campus, focuses mainly on diploma programmes in these areas. Universiti Teknologi Malaysia (UTM) has begun implementing the Quality Assurance schemes in the Faculty of Electrical Engineering at the Skudai campus and the Electrical Engineering department at the Kuala Lumpur campus.

All the engineering faculties at UTM have begun to use the new EAC accreditation requirements that emphasise outcome-based education (OBE) in preparation for the next planned accreditation visit in 2009. In fact, this work begun almost immediately after all the UTM engineering programmes, including those offered by the Faculty of Electrical Engineering at Skudai campus received the full 5 years accreditation from EAC in 2005. However, the accreditation requirements for the last visit were based on the 2004 EAC Manual that did not include a strong emphasis on OBE. However, some of the engineering faculties in the University had already obtained ISO certification for their programmes, which facilitated the development of the procedures and documentations required for the next EAC accreditation in 2009.

The Faculty of Electrical Engineering (FKE) currently offers eight programmes, six of which obtained accreditation from the EAC during the 2005 exercise. Hence, the remaining two programmes will be applying for accreditation in 2009. FKE has taken several important steps in its move toward obtaining full accreditation for all of its current 4-year programmes in 2009. The first is to establish an Academic Quality Committee chaired by the Deputy Dean for Academic and Continuing Education and consisting of ten Head of Departments as well as senior representatives from each department. Another important step was to send the faculty's management team and some senior lecturers for training on OBE. This was part of UTM's capacity building effort to ensure that the process of fulfilling the new EAC accreditation requirements can be carried out smoothly and efficiently. At the same time, courses and workshops related to new teaching and learning methods like critical thinking, active learning, cooperative learning, and problem-based learning were being offered at the university level. UTM's Centre for Teaching and Learning conducts these courses and workshops on a regular basis so that instructors can learn how to use methods aligned with the OBE approach. In fact, as early as 2003 some lecturers in various engineering faculties in UTM have started using new teaching and learning methods in their classes.

Each course[3] at FKE is coordinated by a senior lecturer, who is very familiar with the course requirements based on his or her years of teaching the course. One of the roles of a course coordinator is to make sure that the same quality of teaching occurs in all classes for the same course, which service a range of students from various programmes. The *quality of teaching* refers to not only the course content but also to the standard evaluation tools used such as tests and final examinations.

In relation to programme accreditation, all course coordinators are required to submit detailed course information to the faculty including the list of the course outcomes, the mapping of the course outcomes to the programme outcomes and the evaluation tools used. In this regard, appropriate assessment and evaluation methods have been identified and developed to measure the extent to which intended learning outcomes are achieved both at the course level and programme level. This is essential to support the continual improvement of outcome-based education. It is strongly believed that through the assessment of outcomes, factors such as learning, institutional effectiveness and accountability can be improved (Government of Malaysia 2002). With well-documented results, continuous quality improvement in a programme can be realised, thus *closing the loop* embodied in the outcome-based education approach.

Another important element of the Faculty of Electrical Engineering Quality Assurance initiative is the involvement of the Kuala Lumpur Campus or City Campus diploma curriculum programmes. While engineering diploma programmes do not need to be accredited under EAC, the City Campus diploma curriculum conforms to the knowledge content established by the EAC, since the diploma programmes on the Kuala Lumpur Campus are feeders to the bachelor degree programmes at the Skudai campus. All four City Campus diploma programmes have been approved by the MOHE and also recognised by the Public Services Department (PSD), a government agency responsible for recruiting new civil servants.

The curriculum is reviewed every 3 years to remain current with scientific, technological and knowledge advancement as well as to meet the needs of society at large. Students are provided with written information about the educational goals, course outline and learning outcomes, and the methods of assessment of each

[3]In UTM, a course is equivalent to a content subject. To illustrate, Electronics, Microprocessor and Circuit Theory are examples of courses offered in an Electrical Engineering Program.

programme through e-learning, handouts and academic guides. Lecturers have adopted new teaching and learning methods in some of the subjects. The students are also asked to develop small projects within certain courses, particularly in their final semester. They are encouraged to enter various competitions organised by the City Campus and the national-based challenges in order to gain diverse skills including thinking skill, problem solving and team work.

It is crucial for all the programmes in FKE to obtain full accreditation from the EAC in the next accreditation visit, as well as the future ones. Continuous support from faculty members is necessary to ensure that the quality of the programmes offered is maintained. From a global perspective, success in obtaining full accreditation of the programmes offered at FKE from the EAC will actually put Malaysia a few steps closer to being a full signatory member of the Washington Accords.

The EAC panel itself is under pressure to ensure that the evaluation process is conducted objectively without any prejudice, according to the accreditation criteria stated in the EAC manual. Only then, when all the electrical engineering programmes have obtained full accreditation from the EAC, can the FKE programmes be considered of high quality and be globally accepted.

## Conclusions

Earlier sections in this chapter have given a detailed account of the process of Quality Assurance in Malaysia with emphasis on the engineering programmes. All Malaysian universities are moving toward improving the quality of programmes offered as exemplified by the Faculty of Electrical Engineering case study.

The requirements of both MQA and EAC have pressured the Faculty of Electrical Engineering at the Skudai campus and the Electrical Engineering department at the Kuala Lumpur campus to redesign the structure and curricula of their engineering programmes. This exercise has accelerated the introduction of a new educational philosophy and approaches to teaching and learning among faculty members. Hence, the quality of teaching and learning has shown tremendous improvement. Accreditation is directly related to an improvement in quality due to the rigorous process involved in obtaining it.

Another valuable consequence of this Quality Assurance exercise is the realisation of the need for the collection and safe-keeping of documentation of teaching and learning effectiveness. The requirement for extensive documentation has greatly improved the data collected on the teaching and learning process. Of course, the documentation itself should not be the measure of the quality of the actual teaching and learning. This should be the student learning outcomes that are documented. In fact, the substantial amount of paperwork required can negatively affect faculty members by distracting their focus from their core business of teaching.

Given the existence of a very demanding national Quality Assurance system, MQA and EAC, many Malaysian academic institutions are reluctant to commit to the continuous assessment and improvement of their programmes since this requires a significant level of effort on the part of faculty members and of resources on the part of institutions. In addition, the most trying for the academics is the continual changes in the Quality Assurance processes. With both MQA and EAC operating the Quality Assurance process, some conflicting requirements inevitably exist resulting in redundancies in assessment procedures and additional time-consuming activities, especially for Malaysian universities engaged in engineering programmes. It is hoped that these issues can be resolved as Quality Assurance in Malaysia evolves and matures.

## References

Ariff, M. M. (2007). *The difficult path to education excellence*. Bernama.com. Retrieved November 13, 2007. http://www.bernama.com.my/bernama/v3/news_features.php?id=295773. Retrieved March 24, 2008.

Department of Statistics Malaysia. (2006). *Labour Force Survey 2006*. http://www.statistics.gov.my/eng/

Engineering Accreditation Council. (2007). Engineering programme accreditation manual. Engineering Accreditation Council, Board of Engineers Malaysia. http://www.bem.org.my/v3/downloads.html

Government of Malaysia (2002). *Report for graduate unemployment from public and private tertiary institutions: Causes and recommendations*. Cabinet Committee for Workforce from Public and Private Tertiary Institutions.

International Engineering Alliance. (2007). *International educational accords resulting from IEM Washington (updated Wed 15th Aug 2007): Rules & procedures*. http://www.washingtonaccord.org/Rules-and-Procedures-Aug-2007.pdf. Retrieved March 14, 2008.

Malaysian Qualification Agency. (2008). *MQA in brief*. http://www.lan.gov.my/eng/introduction.cfm. Retrieved March 23, 2008.

Ministry of Higher Education. (2005). *Quality assurance in public universities of Malaysia*. http://www.moe.gov.my/tayang.php?laman=utama&unit=utama&bhs=en

Mohammad, S. (2007). *Outcome based education at UTM*. Workshop on efficiency level assessment course. Skudai, Johor, Malaysia, June 2007.

Patil, A. S., & Pudlowski, Z. J. (2004). *Strategies for international accreditation and recognition of engineering and technology education in the Asia-Pacific Region*. Paper presented at the 4th Global Congress on Engineering Education, Bangkok, Thailand, 5–9 July 2004.

Prados, J. W., Peterson, G. D., & Lattuca, L. R. (2005). Quality assurance of engineering education through accreditation: The impact of Engineering Criteria 2000 and its global influence. *Journal of Engineering Education, 94*(1), 165–183.

Puteh, M. (2006). *Electronic learning in Malaysian Universities and the transformation of Malaysia*. Doctoral Thesis, Monash University, Australia.

Wan Badaruzzaman, W. H. (2005). *Elements of OBE in the accreditation process*. Workshop on continuous quality improvement through outcome based education, Seri Kembangan, Selangor, Malaysia, June 2005.

# Quality Assurance in Engineering Education in the United States

**Lance Schachterle, Chrysanthe Demetry, and John A. Orr**

**Abstract** In USA there is not a federal system of higher education, but a diverse and independent nation-wide set of public and private institutions. Consequently, for the last 100 years, Quality Assurance in US higher education has been performed not by the federal government but by the voluntary regional and disciplinary accrediting agencies such as the Accreditation Board for Engineering and Technology (ABET 2008). Through the involvement of their members these agencies set standards (most recently Engineering Criteria 2000) and then use these standards as part of a peer-review process to accredit engineering education programs. Engineering Criteria (2000) focuses considerable attention on student learning and, while it has been in effect for less than a decade, early studies point to its success in improving engineering education. The experience of Worcester Polytechnic Institute in three ABET accreditation cycles (1996, 2002, 2008) is used to illustrate the impact of *EC2000*.

## Higher Education in the United States: Independence and Diversity

In USA, higher education institutions (HEIs) have from their beginning enjoyed greater self-governance and less central government oversight than in most countries. Three reasons contribute to this condition.

First, from the mid-seventeenth century to mid-nineteenth century, almost all the US HEIs were created by religious communities and governed by local religious boards. The curriculum consisted of classical and religious studies with the primary purpose of the professional training of the clergy (Bishop 1962). Even though many of these early colleges have grown into universities that enroll a diverse

L. Schachterle (✉)
Academic Affairs, Worcester Polytechnic Institute, 100 Institute Road,
Worcester, MA 01609, USA
e-mail: les@wpi.edu

A.S. Patil and P.J. Gray (eds.), *Engineering Education Quality Assurance: A Global Perspective*,
DOI 10.1007/978-1-4419-0555-0_13, 

student body from across USA and around the world, control and, in some sense, ownership has remained *private* with local boards of trustees ardently guarding the independence of their institutions.

Second, as early as 1787 the precedent was set for support of public education by the Continental Congress in the Northwest Ordinance (1787): "Knowledge, being necessary to good government and the happiness of mankind, schools and the means of education shall forever be encouraged." As USA expanded into the Louisiana Territory, acquired by Thomas Jefferson in 1804, over 50% of the US population was employed in agriculture. In addition, in the middle of the 1800s the industrial revolution was mechanizing both industry and agriculture.

In order to prepare the population for these changes, in the latter part of the nineteenth century, state governments, through the support of the Morrill Act, which was signed into law by Abraham Lincoln on July 2, 1862, began to establish public HEIs in the form of *land-grant* colleges. They were called *land-grant* colleges because under the act each eligible state received a total of 30,000 acres of federal land, either within or contiguous to its boundaries, for each member of Congress the state had as of the census of 1860. The purpose of these *public* colleges was (Morrill Act 1862):

> without excluding other scientific and classical studies and including military tactics, to teach such branches of learning as are related to agriculture and the mechanic arts, in such manner as the legislatures of the States may respectively prescribe, in order to promote the liberal and practical education of the industrial classes in the several pursuits and professions in life.

Today most US engineers attain their undergraduate, Bachelors degrees from such public institutions.

Despite the Federal *land-grant* origins, as state colleges and later universities, these institutions were governed by the State legislatures, not by any central authority in the federal government. This is consistent with the United States Constitution's Tenth Amendment that affirms States' rights to organize such areas of social welfare as public education.

Third, American higher education is characterized by a large number of HEIs of various sizes, locations, and missions, all eagerly trying to differentiate themselves from their peers to attract students. The annual *Almanac* edition of The Chronicle of Higher Education (2008, p. 6) reports that there are 2,629 four-year HEI in USA with 1,533 private, 643 public, and 453 proprietary (for profit) institutions. Public and private 2-year institutions, offering Associate's and other professional or technical degrees and certificates, contribute another 1,685 to the rich mix of higher education opportunities available in USA.

The presence of over 4,000 HEIs in USA reflects the history of local and/or state initiatives in establishing colleges and universities. And, while there are several large state systems of higher education (for example, the State University of New York, SUNY, and California State University, CSU), there is no national university system in USA. Many stakeholders in American higher education are persuaded that the competition and resulting innovation fostered by this diversity and independence have resulted in the high quality of American higher education that we see today and, therefore, continue to resist standardization or oversight by the Federal government.[1]

## Quality Assurance of Higher Education in the United States

Given the tradition of independence of higher education in USA, it is not surprising that the process of Quality Assurance is one of independent self-regulation through a peer-review process. Together with this independent tradition the diversity of HEIs has fostered a decentralized, mission-centered approach to Quality Assurance in USA.

### *Regional Accrediting Agencies*

Quality Assurance of US higher education based on a scheme of professional authority gained through experience began with the establishment of the North Central Association of Schools and Colleges, Higher Learning Commission in 1895. Therefore, instead of a nation-wide structure for higher education Quality Assurance, US HEIs have set up voluntary accrediting associations organized on a geographic region basis.

> *Accreditation* is the primary means of assuring and improving the quality of higher education institutions and programs in the United States. Active for the past 100 years, this private, voluntary system of self examination and peer review has been central to the creation of a U.S. higher education enterprise that is outstanding in many respects. (CHEA 2008)[2]

These accreditation regions are defined geographically (New England, Middle States, Southern, North Central, Western, and Northwest) so as to better represent the culture and norms of their members. This regional organization makes it more convenient for members to meet regularly to review standards and policies and to conduct periodic multiday accreditation site visits.

Central to regional accreditation is an institutional self-study and then a campus visit by an external team charged with evaluating the degree to which an institution meets the standards set by its respective accrediting agency. While the language and

---

[1]For in-depth discussions of the history of higher education in the US, see Thelin (2004), Goodchild and Wechlser (1997), Lucas (1994), Rudolph (1962), Veysey (1965), Hofstadter and Hardy (1952), and Knight and Hall (1951).

[2]The Council for Higher Education Accreditation (CHEA) is a national advocate and institutional voice for self-regulation of academic quality through accreditation, CHEA is an association of 3,000 degree-granting colleges and universities and recognizes 60 institutional and programmatic accrediting organizations at http://www.chea.org/. CHEA often takes a leading role in negotiating proposed federal rules changes with the Department of Education.

details of accreditation standards vary by region, all of the standards address such topics as institutional governance, faculty qualifications, student admission procedures, resources allocation and institutional renewal, and, most importantly, the assessment of student learning and institutional effectiveness in accomplishing its missions. At the center of the accreditation process are the educational programs with their claims of what graduates are intended to learn.[3]

After a self-study and site visit from peers, a member institution may be required to take remedial actions and to report formally on how it has come into compliance with the particular standards. Institutions in compliance with the standards of a regional agency are said to be accredited by that agency for a specified time period (usually 10 years), after which another self-study and site visit are required.

## *Federal Legislative Basis for Accreditation and Recent Changes*

The six regional agencies are *recognized* by the US Department of Education to accredit HEIs and, thereby, provide federally sanctioned Quality Assurance as a result of the recently reauthorized *Higher Education Opportunity Act*, *HEOA*, (Public Law 110-315) (HEOA 2008), which was first enacted in 1965. In reauthorizing the *HEOA*, Congress reaffirmed the role of the regional accrediting agencies or associations to set standards of accreditation, including the assessment of institutions' success with respect to student achievement.

However, as described by Judith Eaton, President of the Council for Higher Education Accreditation (CHEA), there are a variety of provisions in the reauthorization that "embody changes in the accreditation–federal relationship and the institution–federal relationship" (Eaton 2008a). In both cases, the changes are driven by (1) new government requirements for accreditation and higher education operation, (2) expansion of areas already under federal examination, and, *crucially*, (3) extensive new federal authority to report data to the public (Eaton 2008a). Eaton (2008a) goes on to point out significant changes in eight accreditation-related areas:

(1) student achievement, (2) appointment of the national advisory committee, (3) due process associated with accreditor review and appeal procedures, (3) institutional mission, (4) distance education, (5) transfer of credit, (6) monitoring of enrollment growth, (7) information to the public and (8) religious mission.

She notes some desirable changes (2008a), for example,

the new law makes it clear, for the first time at the federal level, that institutions play a central leadership role in setting standards and evaluating student achievement. Colleges and universities, not government, determine institutional quality based on judgment of student learning outcomes.

---

[3]The CHEA Web site provides links to regional, disciplinary, and other accreditation groups where descriptions of their standards and accreditation procedures can be found at http://www.chea.org/Directories/index.asp.

However, it is within the area of accountability that there are the most drastic changes to the institutional–federal relationship (Eaton 2008a).

The new law rearranges the institution–federal relationship in two major ways. Similar to what has happened with accreditation, institutions now have (1) a host of new areas of reporting and (2) expanded reporting in areas that are already in the law, culminating in 110 new reporting, record-keeping and regulatory requirements.

Moreover, the new law contains a number of studies that, in many instances, will also require information from colleges and universities and result in additional federal reporting. These include a study of employment upon completion of a program or credential, a study to evaluate the quality of distance education, an examination of proprietary institutions and separate studies of endowments, textbooks, and articulation agreements.

This enlarged platform of information and reporting also significantly expands the potential and capacity of the federal government to embrace practices about which institutions and accreditors continue to have serious concerns. It is an easy step from arraying a bevy of new data to publishing comparability analyses to publishing rankings and to undertaking qualifications comparisons.

Such actions readily lend themselves to standardization and centralization of control of higher education; in short, a national-based system in the making, a system to which, to date, USA has refused to accede. Whatever its shortcomings, the strength of the US higher education enterprise has been vested in its responsible independence, decentralization, and diversity – characteristics associated with institutional leadership and not centralized control.

In a subsequent essay Eaton (2008b) points out that, taken together these changes serve to undermine the traditional model of institutional leadership of higher education that has guided the enterprise for many years. The reauthorized law achieves this through establishing, for the federal government, a new and powerful role of nationwide higher education spokesperson, with the U.S. Department of Education (USDE) functioning as a kind of "Federal Educator-in-Chief."

The primary federal means of enforcing these new accountability requirements is the qualification for a variety of federal tuition and other financial programs.

## History of United States Engineering Education and its Accreditation

Up until the nineteenth century, all the learned professions save theology inducted members through apprenticeship. Doctors, lawyers, and engineers learned their trades following the medieval practice of studying with established professionals. In the early twentieth century, all these professions began to establish more formal expectations for membership, and gradually, a 4-year undergraduate education became expected as the minimum requirement for preparation for professional practice. However, in contrast to other early professional schools that were initially founded independently of colleges and universities, "almost from its beginnings, engineering education in USA was in all essential aspects a form of collegiate education, instituted and directed by educators, rather than by practitioners" (Grayson 1980, p. 373).

The first recorded instance of concern for professional development of engineers in USA was General George Washington's general order in 1778 calling for the establishment of a school of engineering (Grayson 1980). Grayson goes on to point out that (1980, p. 373), "Engineering education in USA began in the early 1800s for the purpose of promoting 'the application of science to the common purposes of life.'" The founding in 1802 of a military academy at West Point, NY, with a strong emphasis on engineering, was an early federal response to this need. And as noted above, these concerns ultimately led to the founding of the *land-grant* institutions.

## *The Evolution of Engineering Education Quality Assurance*

It is within this tradition of independent and educator-run schools that the Engineering Council for Profession Development (ECPD) was founded in 1932. This marked a significant advance in the professionalization of engineering education. In 1980 the council became the Accreditation Board for Engineering and Technology or ABET.[4]

In the period from the founding of ECPD to the 1990s, the measure of quality of engineering programs was the success of their students in taking and passing undergraduate courses in engineering, science, and mathematics. Within the typical 4-year US undergraduate engineering program, which had become *standardized* through the efforts of ECPD, students seeking engineering Bachelors degrees were required to pass at least a year of course work in relevant mathematics and science areas, and a year and half of study in engineering science and design courses. An additional half year of course work in the humanities and social sciences was also mandated. These courses often emphasizing topics of relevance to engineers such as the history of science and technology, professional communications and writing, professional ethics, and engineering economics.[5]

---

[4] See the chapter "Quality Assurance in the Preparation of Technical Professionals: The ABET Perspective" by Peterson, for a detailed discussion of ABET. For a history of ABET, see Prados et al. (2007); for current information, see the ABET Web site: http://www.ABET.org.

[5] It should be noted that course work in the United States is organized by credit hours. These are calculated in term of the hours of classroom and/or laboratory contact that students have with an instructor each week. Therefore, a typical 3 credit-hour course has 3 h of lectures each week and a 4 credit-hour course will add an hour of laboratory contact. Typically US students enroll in 4–6 courses or for 12–18 *credits* per 15-week semester, which amounts to 180–270 contact hours per semester. Over the course of a 4-year, or in some cases 5-year undergraduate career, engineering students accumulate between 120 and 130 credits depending on the ABET course requirements related to a particular field of engineering and other local graduation requirements. Usually students are given grades (*A*, *B*, *C*, *D*, or *F*) at the end of each self-contained course, with overall honors being determined at graduation by the accumulated grade point average (GPA). GPA is calculated by dividing the number of points accumulated ($A=4$, $B=3$, $C=2$, $D=1$, $F=0$)

In the late 1990s, many senior engineering faculty and professionals became increasingly concerned not only with the growing complexity and, at the same time, more narrow focus of the courses needed to graduate from an ABET-accredited program, but also with the increasing pace with which course content became outdated. In the same period, educators in many fields began to recognize that what mattered in education was not what courses students had passed but what they could do with what they learned. Defining such *learning outcomes* – what students could do – was especially important for engineering since as novice professionals students are expected to solve problems and create designs in their field, not simply to show their academic qualifications by continuing to pass courses.

## *Engineering Criteria 2000*

Stimulated by this new emphasis on measuring what engineering graduates could actually do when first on the job or in graduate school, ABET designed an entirely new accreditation protocol (Engineering Criteria 2000 or *EC2000*) for assessing the preparation of candidates for the engineering profession. *EC2000* embodies a set of program standards including student outcomes (a)–(k) that are intended to serve both as Quality Assurance criteria and an impetus to continuous improvement. These criteria addressed all of the significant components of engineering education: faculty qualifications, curriculum structure, student characteristics, support functions such as laboratories, the library and information technology infrastructure, student learning outcomes and objectives, and processes for continuous improvement.[6]

Under *EC2000*, each disciplinary engineering program was required to define what its current graduates should be expected to do, in terms of the program *objectives* (*what recent graduates should be able to do*) and *outcomes* (*what graduating students should be able to do*), as well as have a process for *continuous improvement* based on the assessment of the program outcomes. The relevant Criteria for Accrediting Engineering Programs (ABET 2008, pp. 1–2) include:

### Criterion 2. Program Educational Objectives

Each program for which an institution seeks accreditation or reaccreditation must have in place:

---

by the total number of credit hours taken. A 4.0 GPA is equivalent to straight *As*. This differs from other systems, such as the European Credit Transfer and Accumulation System (ECTS) that bases *credits* on estimates of the hours students spend on coursework per semester. (See the description of ECTS in the chapter "Engineering Education Quality Assurance: The Essential Pillar of Higher Education Reform in Lithuania" by Valiulis and Valiulis.)

[6] See also the ABET site at: http //www.abet.org/forms.shtml#Applicable_to_All_Programs.

(a) Published educational objectives that are consistent with the mission of the institution and these criteria.
(b) A process that periodically documents and demonstrates that the objectives are based on the needs of the program's various constituencies.
(c) An assessment and evaluation process that periodically documents and demonstrates the degree to which these objectives are attained.

## Criterion 3. Program Outcomes

Engineering programs must demonstrate that their students attain the following outcomes:

(a) An ability to apply knowledge of mathematics, science, and engineering
(b) An ability to design and conduct experiments, as well as to analyze and interpret data
(c) An ability to design a system, component, or process to meet desired needs within realistic constraints such as economic, environmental, social, political, ethical, health and safety, manufacturability, and sustainability
(d) An ability to function on multidisciplinary teams
(e) An ability to identify, formulate, and solve engineering problems
(f) An understanding of professional and ethical responsibility
(g) An ability to communicate effectively
(h) The broad education necessary to understand the impact of engineering solutions in a global, economic, environmental, and societal context
(i) A recognition of the need for, and an ability to engage in life-long learning
(j) A knowledge of contemporary issues
(k) An ability to use the techniques, skills, and modern engineering tools necessary for engineering practice.

Program outcomes are outcomes (a)–(k) plus any additional outcomes that may be articulated by the program. Program outcomes must foster attainment of program educational objectives. There must be an assessment and evaluation process that periodically documents and demonstrates the degree to which the program outcomes are attained.

## Criterion 4. Continuous Improvement

Each program must show evidence of actions to improve the program. These actions should be based on available information, such as results from Criteria 2 and 3 processes.

To respond to the growing national movement for greater transparency intended to enable students and parents and the public in general to understand the claims for success of each engineering program, every program needs to show how its *outcomes* and *objectives* are related to the overall *outcomes* and *objectives* of the institution of which they were a part. Furthermore, the *outcomes* and *objectives* need to be defined and continually reviewed through the on-going advice of the *stakeholders* in higher education for engineering. These stakeholders included not only faculty members, students, and their tuition paying parents, but also state legislators, alumni, professionals in practice, and representatives of companies hiring graduates. This is based on the assumption that gathering the perspectives of a variety of stakeholders will enhance aware of the ever-changing expectations for engineering practice and, thereby, provide guidance for curricular improvements.

Just as the *continuous improvement* has swept global business world, continuous program improvement based on documenting and correcting inadequacies in learning *outcomes* and *objectives* (*closing the loop,* in ABET jargon) has now became central to higher education and engineering education *Quality Assurance* processes.

## New Ways to Assess Learning Outcomes

Engineering program faculties now have to rethink how they know what students were actually learning. Transcripts of courses passed can no longer be used to support claims of student learning – no surprise really to most faculty and students!

### *Assessing Outcomes*

Such a wholesale rethinking of how to assess the success of engineering programs is a very complex and very time-consuming task. Nor is it a task that all faculty members were equally prepared to perform. In response to these concerns, some engineering institutions hired experts with social science training in the areas of curricular innovation and measuring learning outcomes to design their assessment processes. Other institutions struggled, with mixed success, to find engineering faculty members willing to undertake these new challenges, believing that the faculty as a whole is more likely to accept a system to assess learning outcomes that was devised by their colleagues, rather than one imported from the social sciences.

ABET assisted institutions to adapt these new standards embodied in *EC2000* by providing seminars and advice by Gloria Rogers, ABET's Associate Executive Director of Professional Services. Dr. Rogers, who had a well-established career in assessment of learning outcomes at the Rose-Hulman Institute of Technology, authored a number of influential papers and guides, and conducted annual meetings to help her colleagues understand and implement these radically new expectations.[7]

---

[7] See Rogers (2000) and http://www.ABET.org/assessment.shtml for current information. Additional publications related to outcomes assessment in engineering education by Rogers and others can be found in the American Society for Engineering Education Annual Conference Proceedings: http://www.asee.org/conferences/v2search.cfm; Frontiers in Education Conference Proceedings: http://www.fie-conference.org; *International Journal of Engineering Education*: http://www.ijee.dit.ie; and *Journal of Professional Issues in Engineering Education and Practice*: http://pubs.asce.org/journals/professionalissues/.

As additional assistance, other engineering education scholars have described outcomes assessment from a systems perspective, have proposed methods to interpret outcomes defined in Criterion 3, and have suggested way to create and disseminate research-based assessment tools (Olds and Miller 1998; Besterfield-Sacre et al. 2000, 2002; McGourty et al. 2002; Olds et al. 2005).

## *The Capstone Design Experience*

One of the first responses to these new ABET expectations for assuring the quality of learning outcomes was increased emphasis, usually in the senior year, of a *capstone design experience*. ABET anticipated that such a design experience would require students to synthesize their previous learning in mathematics, science, and engineering science and design in order to address a problem that they might reasonably be expected to encounter in their first year of professional practice. Engineering institutions thus quickly moved to create opportunities, often embedded in senior-year disciplinary seminars, for students to demonstrate this ability to solve problems. Examination of design projects and capstone design experiences has become a major element of the outcomes assessment process in virtually all engineering programs (Napper and Hale 1999; McKenzie et al. 2004).

## *Linking Courses to Learning Outcomes*

While opportunities to learn teamwork and carry out projects began to suffuse the whole curriculum, engineering students still spent most of their time in classroom settings. Faculty thus began to link demonstrations of at least minimum competence needed to pass selected courses to a subset of their programmatic learning outcomes. That is, course objectives were often reexamined and clarified in the context of ABET's Criterion 3 outcomes (Felder and Brent 2003; Howell et al. 2003). For specific courses, faculty defined specific learning outcomes in such a way that unless students satisfied those outcomes, they could not pass the course. By having a multiplicity of courses which, at various levels and in various ways, challenge the students to meet all the programmatic learning outcomes, faculty could still determine how well students were achieving overall learning outcomes through course work. Mapping outcomes to courses and then relating them to embedded assessment methods such as exams, quizzes, tests, problem sets, assignments, and class projects provide a way to gather direct evidence of student learning, a critical aspect of *Quality Assurance*, that is a natural part of the educational process and not an added burden on faculty or students.

## *The Professional Engineer Certification*

In USA, professional licensing (which grants the title of Professional Engineer, P.E.) is easier to obtain for students who have graduated from ABET-accredited programs. (Requirements for professional licensing are established in USA at the state level and thus vary significantly from state to state). Such professional licensing is most important in those areas of engineering requiring governmental approval of construction drawings and similar documents by a registered professional engineer (P.E.). Consequently, in USA most civil engineers seek P.E. status. However, unlike in other counties, comparatively few engineers outside of civil engineering seek the additional status of the P.E. designation. The pass-rate on P.E. state examinations is one external measure of Quality Assurance in an engineering program, but in the absence of a national exam across all states and all disciplines, it is of limited value.

Nevertheless, some engineering programs are making use of information on student performance from the Fundamentals of Engineering Exam (FE), the first step in seeking licensure, as one element of their outcomes assessment process (Nirmalakhandan et al. 2004; Koehn et al. 2008). The FE is administered by The National Council of Examiners for Engineering and Surveying, NCEES (2008) and covers "subject matter taught in a typical EAC/ABET-accredited baccalaureate engineering program. It appropriately covers a comprehensive range of subjects in engineering."[8]

Nevertheless, for most engineering programs, ABET accreditation is the primary *Quality Assurance* mechanism relied on to guarantee acceptable levels of performance of undergraduates related to the expectations of the engineering profession.

## *Subjective Evidence from Graduating Students and Alumni/ae*

Two types of survey instruments have been developed to provide information on learning outcomes from graduating students and engineering program alums. Such surveys are limited by the necessarily subjective nature of student opinions of their own achievement. However, if the survey results represent a significant numbers of respondents, then individual programs can be relatively confident in their reliability. In addition, for national surveys, if responses are obtained from a broad range of graduates from a variety of programs and institutions then the overall and comparative data may be of some value.

[8] The National Council of Examiners for Engineering and Surveying (NCEES) is a national nonprofit organization composed of engineering and surveying licensing boards representing all states and US territories. NCEES develops, scores, and administers the examinations used for engineering and surveying licensure throughout the United States. NCEES also provides services facilitating professional mobility for licensed engineers and surveyors. NCEES is an accredited standards developer with the American National Standards Institute (ANSI). Information about the FE exam can be found at: http://www.ncees.org/exams/fundamentals/.

Engineering programs often use one or both of the following national surveys: the Educational Benchmarking, Inc. (EBI) and the National Survey of Student Engagement (NSSE).

The EBI Engineering Education Assessments (Educational Benchmarking, Inc. 2008):

measure the effectiveness of your program from your student's perspective. You will learn what key dimensions of your program are the strongest and which areas need to be improved. The knowledge you will gain from our assessments will drive and sustain your continuous improvement program.

The EBI Engineering Education Assessments offers three surveys, an Exit Assessment, an Alumni Assessment, and an Employer Assessment.[9] Engineering program most often use the exit assessment as a means of assessing programmatic strengths and weaknesses and thus directing *Quality Assurance*.

The National Survey of Student Engagement (NSSE pronounced *Nessie*) is a national survey of freshman (entering students) and seniors (graduating students) administered at the institutional level. The NSSE gathers information on students' perception of their learning and the classroom and institutional activities and expectations that promote learning. Institutions are encouraged to use their results as a way to provide greater transparency about institutional quality as well as guide improvement.[10] In this regard, the NSSE is seen as an alternative to the *US News & World Report, Time Higher Education* and other rankings that are based on higher education peer ratings of institutional resources and reputation. Instead, the EBI and NSSE encourage participating programs and institutions to identify peer groups with whom they wish to compare their results as a means of *benchmarking*. They also provide very detailed reports that assist in the interpretation of results for internal and external audiences and to use the results for improvement.

Many institutions also solicit information on job satisfaction and advancement from their graduates or their employers (Puerzer and Rooney 2002; Koehn and Parthasarathy 2005), but tracking graduates is notoriously difficult and statistically useful samples are hard to obtain. And, data from graduates themselves are subjective and often colored by strong feelings of loyalty (or animosity). In addition, employers are reluctant to share information about employees. As a result, such information has to be interpreted carefully and used in conjunction with as much other information as possible in order to come to reasonable conclusions about a program's quality.

## *Three Cycles of Accreditation at WPI under EC2000*

Readers may get some sense of the progress made over the 20 years in developing Quality Assurance programs that are based on measuring learning outcomes, by considering the case of Worcester Polytechnic Institute (WPI). In the late 1960s,

[9] See the EBI Engineering Education Assessments Web site for descriptions of these surveys: http://www.webebi.com/_AsmtServices/Engineering/default.aspx.

[10] See the NSSE Web site for additional information: http://nsse.iub.edu/index.cfm.

the engineering faculty at WPI became concerned with the inadequacies of the conventional course-based educational program, which WPI had provided since its founding in 1865. WPI faculty members were among the first to recognize that measuring student achievement and the potential for professional success by looking only at what courses students passed was of very limited value. Real engineers solve real problems rather than sit in classrooms.

Thus, WPI's new academic program, implemented between 1970 and 1972, relied more on the demonstration of student ability through project work and other student accomplishments and correspondingly less on passing courses. The new academic program had at its center two measurements of student achievement, namely, a capstone design project of significant scope (at least a quarter of a year of work, usually in the senior year) and a competency examination, an open-ended, multiday examination which students could take on several occasions in their senior year.

The competency examination ultimately proved too difficult to implement consistently and fairly, and was dropped in 1984 in favor of a set of course requirements. This change, not accidentally, corresponded to the revision of ABET's requirements at the same time.

The senior design project, as well as a junior year interdisciplinary project carried out abroad by many students, remained as hallmarks of the assessment process at WPI. There are three advantages of these projects. The first is their comparatively large scope, which mimics those in the *real world*; second, is their emphasis on teamwork, a very important aspect of professional competence; and, third, is the expectation that students will address problems (often provided by professional partners) that are similar to those that they may encounter early in their careers. WPI faculty reviewed the senior year projects in order to assess how well students were achieving desired outcomes and to identify what pedagogical improvements in course work were required to enhance students' performance in subsequent senior year projects (Schachterle 1998).

Based on our previous experience with such authentic assessment methods as the student projects, when *EC2000* was being finalized in 1996 WPI became one of the first two institutions to volunteer to undergo reaccreditation of all its engineering programs using the new criteria. ABET accreditation is on a 6-year cycle. Our primary evidence for learning was the faculty-reviewed assessment by discipline of the major design projects; as an experimental program, WPI's engineering programs were reaccredited for 6 years.

In 2002, WPI went through the reaccreditation process again, using additional data which had been developed since the 1996 visit. Examples of courses that had been redesigned to assure that students achieve a subset of the departmental learning outcomes were added to the emphasis on project work. In 2002, we also documented the ways in which departmental outcomes were correlated with overall university goals. In addition, we provided evidence of our continuous internal and external review of courses and modification of courses wherever necessary. The external review of assessment results was conducted by advisory committees consisting of both WPI graduates and nongraduates, i.e., other professional engineers and employers.

To ensure that faculty members and administrators stay focused on *Quality Assurance*, the faculty voted to create a subcommittee of the WPI Faculty

Governance Committee on Academic Policy. The subcommittee meets regularly to review assessment instruments, conduct assessments, evaluate resulting data, and identify strengths as well as weakness within the curriculum that require attention. The subcommittee has, for example, recommended a revised set of learning outcomes for design projects and has noted concerns from seniors about the effectiveness of some mathematics courses.

In 2008, WPI again prepared for reaccreditation with a site visit planned for November 2008. Central to the 2008, self-study and team visit is a detailed account of continuous improvement efforts since the 2002 reaccreditation. In particular, the self-study report identified the problems that internal and external reviewers have detected related to the achievement of desired student learning outcomes, the curricular improvements that have been made in response to those problems, and the evidence that the problems have been successfully addressed.

A full range of data will be shared with the ABET visiting team. These data include course-based assessment results that are linked to program learning outcomes, internal peer reviews including student and faculty assessments of both written reports and oral presentations related to the senior-year design projects and the junior-year interdisciplinary projects, the EBI and NSSE survey results over several years, alumni/ae survey summaries, and internal senior exit survey findings.

The NSSE review will be especially valuable because WPI organized a *consortium* of peer institutions in 2000. Together the consortium members added a series of questions related to the ABET learning outcomes (a)–(k) to the basic NSSE survey items. These additional items will be used as bench marks by the consortium institutions. That is, having common items will allow consortium members to compare their results related to important engineering learning issues with those of other similar institutions.

WPI has designed a matrix that indicates how our learning outcomes are assessed by various measurements, as have other HEIs. This matrix may be viewed at the home page of the Assessment Plan for Institutional Learning Outcomes.[11] This site also contains Quality Assurance documents from the 2002 ABET visit as well as the results of the recent reaccreditation by our regional accrediting agency, the New England Association of Schools and Colleges (NEASC). Material from the 2008 ABET cycle will be posted in 2009, once the final report has been submitted by the visiting team and a decision on reaccreditation has been made.

## Conclusions

If asked how well *EC2000* has succeeded in improving *Quality Assurance* in USA, one might be tempted to respond as Chinese premier Chou En-lai (Zhou Enlai) is alleged to have when asked about the long-term effects of the French Revolution, "it's too early to tell."

[11]The WPI Assessment Plan for Institutional Learning Outcomes Web site is located at http://www.wpi.edu/Academics/Outcomes/.

While it is safe to say that nearly all ABET accredited programs now have in place systems for gathering the data needed to guide continuous quality improvement consistent with the requirements of *EC2000*, several concerns remain. Many institutions and their engineering programs are concerned that the systems they have developed are *overengineered*, i.e., *too many data* are collected from which *too little information* is extracted. Another concern is the difficulty that programs have in demonstrating that a curricular change has addressed an identified problem. This may be due to a lack of sensitivity of the *diagnostics*, i.e., measures may not be sensitive enough to gage the change. Or it may be that the time-line for judging the success of the change is inadequate. Some changes may take years to show results. Also, there is still work to be done to insure consistency among the ABET visiting teams so that the accreditation reports are comparable in thoroughness and quality.

*EC2000* with its emphasis on providing evidence of student learning outcomes has significantly raised the bar for accreditation. While many engineers agree that these aspirational goals for what graduates can actually do are important, assessing them is inherently more difficult than the system that *EC2000* replaced where the criterion was clear if simplistic, courses students passed. Nonetheless, the engineering education literature now includes quite a few examples of positive *lessons learned* as a result of implementing *EC2000* (McGourty et al. 1998; Dabney Creighton et al. 2001; Miller and Olds 2002; Soundarajan 2002).

Of special interest is a nationwide study that investigated the effects of *EC2000* on curriculum and instruction, faculty culture, and perceptions of reward systems. The results of this study are quite positive, suggesting greater curricular emphasis on knowledge and skills aligned with *EC2000* learning outcomes, high levels of faculty buy-in and participation in program assessment, and modest increases in professional development related to undergraduate engineering education (Prados et al. 2005; Lattuca et al. 2006a). As reported by Lattuca et al. (2006b):

Compared to their 1994 counterparts, 2004 graduates report higher ability levels on nine measures of learning. Outcomes are linked to changes in program curricula, instruction, faculty culture, administrative practices and student experiences.

Many observers agree that *EC2000* has sharply refocused faculty members' attention on how well students are learning and how faculty members can gather direct and indirect evidence of student learning from students themselves as well as graduates and other relevant stakeholders. Equally important is the focus on using evidence of student learning to close the loop of assessment, i.e., to guide the continuous improvement of teaching and learning. These efforts have made the outcomes assessment processes required by ABET and the US regional accrediting agencies a part of the normal workload for every accredited engineering department and HEI. Gone are the days when an accreditation self-study can be feverishly assembled by a single administrator a few months before a team visit. In this regard, perhaps the most significant and long-lasting contribution of *EC2000* to engineering education *Quality Assurance* is that it has created a much wider base of stakeholders who are concerned, as part of their professional responsibilities, with the continuous assessment and improvement of engineering education.

## References

ABET. (2008). *Criteria for accrediting engineering programs: Effective for evaluations during the 2009–2010 accreditation cycle*. Retrieved December 2008, from http://www.abet.org/Linked%20Documents-UPDATE/Criteria%20and%20PP/E001%2009-10%20EAC%20Criteria%2011-20-08.pdf

Besterfield-Sacre, M., Shuman, L. J., & Wolfe, H. (2002). Modeling undergraduate engineering outcomes. *International Journal of Engineering Education, 18*(2), 128–139.

Besterfield-Sacre, M., Shuman, L. J., Wolfe, H., Atman, C. J., McGourty, J., Miller, R. L., et al. (2000). Defining the outcomes: A framework for EC-2000. *IEEE Transactions on Education, 43*(2), 100–110.

Bishop, M. (1962). *A history of Cornell*. Ithaca, NY: Cornell University Press.

Council for Higher Education Accreditation. (2008). *Talking points: Accreditation, students and society*. Washington, DC: Council for Higher Education Accreditation. Retrieved December 2008, from http://www.chea.org/pdf/2008_Talking_Points.pdf

Dabney Creighton, S., Johnson, R. L., Penny, J., & Ernst, E. (2001). A comprehensive system for student and program assessment: Lessons learned. *International Journal of Engineering Education, 17*(1), 81–88.

Eaton, J. S. (2008a). The Higher Education Opportunity Act of 2008: What does it mean and what does it do? (First of two articles). *Inside Accreditation with the President of CHEA, 4*(1). Washington, DC: Council for Higher Education Accreditation. Retrieved December 2008, from http://www.chea.org/ia/IA_2008.10.30.html

Eaton, J. S. (2008b). The Secretary shall report to the public: Meet the new nationwide spokesperson – The Federal Educator-in-Chief. (More on the Higher Education Opportunity Act of 2008, Second of two articles). *Inside Accreditation with the President of CHEA, 4*(2). Washington, DC: Council for Higher Education Accreditation. Retrieved December 2008, from http://www.chea.org/ia/IA_2008.11.14.html

Educational Benchmarking, Inc. (2008). *Engineering education assessments*. Accessed November 29, 2008, from http://www.webebi.com/_AsmtServices/Engineering/default.aspx

Felder, R. M., & Brent, R. (2003). Designing and teaching courses to satisfy the ABET engineering criteria. *Journal of Engineering Education, 92*(1), 7–25.

Goodchild, L. F., & Wechlser, H. S. (eds). (1997). *The history of higher education* (ASHE Reader Series 2nd ed.). New York, NY: Simon & Schuster Custom.

Grayson, L. P. (1980). A brief history of engineering education in the United States. *IEEE Transactions on Aerospace and Electrical Systems, AES-16*, 3. Retrieved December 2008, from http://ieeexplore.ieee.org/stamp/stamp.jsp?arnumber=4102331&isnumber=4102314

Grayson, L. P. (1993). *The making of an engineer: An illustrated history of engineering education in the United States and Canada*. New York, NY: Wiley.

HEOA. (2008). The Higher Education Act of (HEOA), Public Law 110–315, enacted on August 14, 2008; a reauthorization of the Higher Education Act of 1965, as amended (HEA). Retrieved December 2008, US Department of Education Web site, http://www.ed.gov/policy/highered/leg/hea08/index.html; http://frwebgate.access.gpo.gov/cgi-bin/getdoc.cgi?dbname=110_cong_public_laws&docid=f:publ315.110.pdf

Hofstadter, R., & Hardy, C. D. (1952). *The development and scope of higher education in the United States*. New York, NY: Columbia University Press.

Howell, L. L., Roach, G. M., Clark, D. C., & Cox, J. J. (2003). Use of explicit instructional objectives to achieve program outcomes and facilitate assessment: A case study. *International Journal of Engineering Education, 19*(6), 828–35.

Knight, E. W., & Hall, C. L. (1951). *Readings in American educational history*. New York, NY: Appleton-Century-Crofts.

Koehn, E., Koehn, J. F., Malani, R. D., & Mandalika, R. (2008). Outcome assessment of performance on the fundamentals of engineering (FE) examination. *Journal of Professional Issues in*

*Engineering Education and Practice, 134*(1), 1–6. Retrieved December 2008, from http://dx.doi.org/10.1061/(ASCE)1052-3928(2008)134:1(1).

Koehn, E., & Parthasarathy, M. S. (2005). Practitioner and employer assessment of ABET outcome criteria. *Journal of Professional Issues in Engineering Education and Practice, 131*(4), 231–237. Retrieved December 2008, from http://dx.doi.org/10.1061/(ASCE)1052-3928(2005)131:4(231).

Lattuca, L., Terenzine, P. T., & Volkwein, J. F. (2006a). Engineering change: A study of the impact of EC2000, full report. *ABET Publication Code 06-ECF*. Philadelphia PA: ABET.

Lattuca, L., Terenzine, P. T., & Volkwein, J. F. (2006b). Panel – Engineering change: Findings from a study of the impact of EC2000, Session M2B. *IEEE 2006 Frontiers in Education Conference*. Retrieved December 2008, from http://fie.engrng.pitt.edu/fie2006/; 1627.pdf

Lucas, C. J. (1994). *American higher education: A history*. New York, NY: St. Martin's Press.

McGourty, J., Sebastian, C., & Swart, W. (1998). Developing a comprehensive assessment program for engineering education. *Journal of Engineering Education, 87*(4), 355–361.

McGourty, J., Shuman, L., Besterfield-Sacre, M., Atman, C., Miller, R., Olds, B., et al. (2002). Preparing for ABET EC 2000: Research-based assessment methods and processes. *International Journal of Engineering Education, 18*(2), 157–167.

McKenzie, L. J., Trevisan, M. S., Davis, D. C., Beyerlein, S. W. (2004). Capstone design courses and assessment: A national study. *Proceedings of the 2004 ASEE Annual Conference and Exposition*. Retrieved December 2008, http://www.asee.org/acPapers/2004-644_Final.pdf

Miller, R. L., & Olds, B. M. (2002). Lessons learned in developing and implementing a program assessment plan. *International Journal of Engineering Education, 18*(2), 217–224.

Morrill Act. (1862). Act of July 2, 1862, Public Law 37-108, which established land grant colleges; Enrolled Acts and Resolutions of Congress, 1789–1996; Record Group 11; General Records of the United States Government; National Archives. Retrieved December 2008, from http://www.ourdocuments.gov/doc.php?flash=false&doc=33#

Napper, S. A., & Hale, P. N., Jr. (1999). Using design projects for program assessment. *Journal of Engineering Education, 88*(2), 169–172.

National Council of Examiners for Engineering and Surveying. Accessed 29 November 2008, from NCEES (2008). http://www.ncees.org/exams/fundamentals/

Nirmalakhandan, N., Daniel, D., & White, K. (2004). Use of subject-specific FE exam results in outcomes assessment. *Journal of Engineering Education, 93*(1), 73–77.

Northwest Ordinance, July 13, 1787; (National Archives Microfilm Publication M332, roll 9); Miscellaneous Papers of the Continental Congress, 1774-1789; Records of the Continental and Confederation Congresses and the Constitutional Convention, 1774-1789, Record Group 360; National Archives; Retrieved December 2008, from http://www.ourdocuments.gov/doc.php?doc=8

Olds, B. M., & Miller, R. L. (1998). An assessment matrix for evaluating engineering programs. *Journal of Engineering Education, 87*(2), 173–178.

Olds, B. M., Moskal, B. M., & Miller, R. L. (2005). Assessment in engineering education: Evolution, approaches and future collaborations. *Journal of Engineering Education, 94*(1), 13–25.

Prados, J. W., Peterson, G. D., & Lattuca, L. R. (2005). Quality assurance of engineering education through accreditation: The impact of Engineering Criteria 2000 and its global influence. *Journal of Engineering Education, 94*(1), 165–183.

Puerzer, R. J., & Rooney, D. M. (2002). The alumni survey as an effective assessment tool for small engineering programs. *Journal of Engineering Education, 91*(1), 109–116.

Rogers, G. M. (2000). EC2000 and measurement: How much precision is enough? *Journal of Engineering Education, 89*(2), 161–165.

Rudolph, F. (1962). *The American college and university: A history*. New York, NY: Alfred A. Knopf.

Schachterle, L. (1998). Outcomes assessment at WPI: A pilot accreditation visit under engineering criteria 2000. *Journal of Engineering Education, 87*(2), 115–120.

Soundarajan, N. (2002). Preparing for accreditation under EC 2000: An experience report. *Journal of Engineering Education, 91*(1), 117–123.

The Chronicle of Higher Education. (2008). The Chronicle Almanac, 2007–8: The Nation. *The Chronicle of Higher Education, 6*, 29 August 2008. Retrieved December 2008, from http://chronicle.com/weekly/almanac/2007/nation/0101201.htm

Thelin, J. R. (2004). *A history of American higher education*. Baltimore, MD: The Johns Hopkins University Press.

Veysey, L. R. (1965). *The emergence of the American University*. Chicago, IL: University of Chicago Press.

# Quality Assurance in Engineering Education: An All-round Perspective

**Kin Wai Michael Siu**

**Abstract** This chapter discusses Quality Assurance in engineering education from an overall perspective. It advocates that engineering curricula should be revised to nurture all-round engineering designers, so that they can meet new social, industrial and educational needs. The chapter reviews the limitations of current engineering education practices, and argues that they are biased either toward the acquisition of engineering and technological knowledge and skills at the expense of critical thinking skills or toward the nurturing of creative thinking, while ignoring the ability to conduct in-depth investigations. This chapter uses Hong Kong as a case study to further identify new social, industrial and educational needs. To meet these needs, the chapter proposes Eight Cs as evaluative criteria for all-around engineering curricula. They are: competent, comprehensive, critical, creative, curious, continuous, collaborative, and compulsory.

## Introduction

Cities such as Hong Kong, together with their educational structures, necessarily react in a dynamic manner to social, economic and technological changes taking place in other parts of the world. For example, within a mere one and a half centuries, the city of Hong Kong has changed from being a fishing port, to an entrepôt, to a manufacturing-oriented economy, to a combination of manufacturing and service industries and, finally, to the international financial centre it is today. Several policy addresses of the city, for example, *The 1997 Policy Address* (Hong Kong Special Administrative Region Government 1997) have described Hong Kong's education policies, including those on curriculum development and assessment. These policies have also changed over time in order to meet the needs of society as it strives to keep pace with the rest of the world.

K.W.M. Siu (✉)
School of Design, The Hong Kong Polytechnic University, Hunghom, Kowloon, Hong Kong
e-mail: m.siu@polyu.edu.hk

A.S. Patil and P.J. Gray (eds.), *Engineering Education Quality Assurance: A Global Perspective*,
DOI 10.1007/978-1-4419-0555-0_14, © Springer Science+Business Media, LLC 2009

Along with the changes in society, engineering curricula have also changed. Given the critical transformations that we face today, it may be well to investigate emerging social, industrial and educational needs, and then to consider how the development and assessment of engineering curricula can evolve to meet these needs.

## New Social, Industrial and Educational Needs and Limitations in Current Curricula

Over the last 10 years, studies have examined the relationship between the technology curricula at the secondary level and engineering curricula at the tertiary level, and the job requirements of engineers in Hong Kong and several other Asian cities (Siu 1997, 2000a, 2002, 2007). The findings indicate that while there have been major changes in the job requirements for engineers, the curricula have not kept pace.

In the past, engineers focused on generating solutions for well-defined problems that were presented to them. Today, the ability to solve problems or follow pre-determined procedures to finish tasks is not sufficient (Starko 2000). Engineers are often required to initiate directions for design, production and management, as well as to make decisions, even when they are not working at the supervisory level. However, studies found that the current engineering (tertiary) and technology (secondary) curricula have paid relatively little attention to enhancing students' decision-making abilities, in particular with respect to *initiative*. Although policymakers and curriculum planners emphasise the importance of providing students with the opportunity to be more creative, the focus of curricula is still mainly on training them to generate solutions for pre-determined problems (sometimes using well-defined model answers). Such a focus will not prepare graduates to meet the job requirements for today's and tomorrow's engineers.

Teamwork is another job skill as well as an aspect of engineering that is being emphasised more and more. Unlike in the past, people in workplaces today are increasingly required to collaborate with others from both their own and different disciplines. Unfortunately, there is little emphasis on teamwork in the secondary curricula, and although there has been a slight improvement in the past few years at the university level, teamwork training and experience are still limited.

It is interesting to note that in the past few decades, more and more females have entered the engineering disciplines and professions and have risen to positions of influence. Increasingly, women are occupying leading positions in international engineering companies, for example, in information technology. However, it is still a fact that young women in many places, particularly in Asian cities, have limited opportunities to study engineering and technology subjects.

This is especially the case in secondary school, which has a ripple effect at the university level. In Hong Kong's secondary technology studies, for example, due to the physical limitations, technology workshops and laboratories can accommodate

20 or fewer students. Since the average number of students in a secondary class is about 40, school administrators usually split classes into two, and have one of the groups attend a technology subject while the other group attends a home economics subject. This split is always according to gender (the so-called rational way) instead of student interest or other considerations. Although in recent years this situation has improved somewhat, it is still a small number of girls that have the opportunity to attend secondary technology classes (Siu 1997, 2007).

A similar situation exists regarding students with special needs. Although the need to provide equal opportunities to learn engineering for students with special needs has been recognised for many years, policymakers have not acted to adapt the engineering curricula for such students.

## Curriculum Development and the Eight Cs

There are many attributes that should be considered by curriculum developers and evaluators as they strive to meet new social, industrial and educational needs. The following Eight Cs are the attributes or touchstones of excellence that can help design and evaluate an all-round curriculum for engineering education. The Eight Cs, are: *competent*, *comprehensive*, *critical*, *creative*, *curious*, *collaborative*, *continuous*, and *compulsory*.

### *Competent*

Engineering education emphasises both the understanding and the application of knowledge. Thus, curricula should allow students to acquire knowledge in different areas related to engineering and technology, so that they have an appropriate context in which to identify and solve problems. For example, they should study different social, cultural, scientific, design, and technology theories. Students should also have the opportunity to apply the knowledge they have acquired using different kinds of tools, methods and techniques (hand tools, machine tools, internet and information tools, thinking tools, etc.). In other words, the engineering curricula should prepare students to be *competent* in handling tools and techniques in order to be able to apply their knowledge and understanding. Project work has been shown to be a good way to motivate students to gain the knowledge they need and to improve their ability to apply that knowledge (Siu 2002, 2007).

### *Comprehensive*

Many years ago, in the days of the apprentice or craft approach to preparing engineering students, trainees were required to be competent in only a few areas. At that time, competence in a particular and often very narrow area of expertise generally secured an engineering graduate a job for life. Even in the 1980s this situation existed in many places around the world. However, today, scientific, engineering and technological developments and consequent requirements are very diverse. Therefore, engineering curricula must be widened to include not just ever-expanding knowledge and abilities, but also a greater array of local and global issues. Students are now expected to explore many areas of interest. Thus, the engineering curricula must be organised to allow students to gain a more *comprehensive* understanding of engineering and related subject areas.

## *Critical*

Engineering curricula should not merely aim at training students to follow fixed or well-defined routes/procedures to tackle pre-determined tasks and to produce *model answers*. Instead, engineering curricula should encourage students to identify problems, to conduct explorations, investigations and analyses, to provide solutions that are not yet identified or pre-determined, and then to carry out evaluations. That is, students should be taught to use *critical thinking skills* for problem solving. Einstein (1938, p. 92) asserted the importance of initiation and problem-finding skills:

The formulation of a problem is often more essential than its solution. ...To raise new questions, new possibilities, to regard old problems from a new angle, requires imagination and marks real advance in science.

Also, Wertheimer (1959, p. 123) pointed out that:

The function of thinking is not just solving an actual problem but discovering, envisaging, going into deeper questions. ...Envisaging, putting the productive question is often a more important, often a greater achievement than the solution of a set question.

Thus, curricula should allow students the opportunity to initiate, explore, identify, make decisions and then present their own opinions and ideas. In other words, nourishing students' critical thinking ability should be an important objective of engineering education. Critical thinking skills come from students' direct experience with engineering problems. Accordingly, a project-based learning and a student-centred approach should be integral elements of engineering curricula.

## *Creative*

Engineering curricula should not aim at training students to be skilled labourers. Today, machine automation processes have replaced many of the routine procedures and processes that previously had to be carried out by hand. In fact, machines perform better than people in terms of quantity of output and levels of quality. Precise and complicated calculations can also be handled by computers. Only people can contribute their minds – their ideas. More precisely, in order to serve society in the future, it is important for students to learn how to generate *creative* ideas rather than model answers.

The literature on creativity is substantial. Many people have analysed creativity to determine what makes a person creative (Mendelowitz 1981; Parnes and Harding 1962; Siu 2007). While the exact nature of creativity remains elusive, there is general agreement that creative people tend to possess certain traits, which include sensitivity, flexibility, originality, playfulness, productivity, fluency as well as analytical and organisational skills (Gilbert 1998; Siu 2002). Thus, engineering curricula should have different kinds of activities to encourage students to develop these traits.

## *Curious*

Although good teachers and physically favourable settings are necessary and desirable, students' curiosity is a crucial factor in their becoming successful critical and creative thinkers. Being curious relates directly to one's interest and motivation. This means that success in engineering is closely related to whether students are interested and motivated in their studies. Therefore, engineering curricula and their related activities and materials should encourage students to be self-motivated and self-initiated to study, to discover, and to explore.

## *Collaborative*

Since today's multi-faceted problems cannot be tackled by one person alone, engineers carry out research and design projects in teams. Therefore, students need to be *collaborative* and to learn teamwork. The ability to communicate is a fundamental skill for collaboration, since it allows students to present what they are thinking to others (Siu 2007). Therefore, the engineering curricula should provide more opportunities for students to collaborate, not only with those studying the same discipline, but also with those from other disciplines and other levels.

There are some joint programmes currently offered by different disciplines, faculties and departments that allow engineering students to collaborate with students from other areas (Siu 2002, 2007). Through collaborative activities that differ in terms of the number of students involved, topics, disciplines, duration, nature and difficulty of tasks, etc., engineering students can discover their strengths, remedy their weaknesses and consequently enhance their future studies and careers.

## *Continuous*

We are living in a constantly changing world where change is more rapid than ever before. Thus, engineering curricula face pressure for continuous improvement. Curricula should be constantly reviewed and updated. They should be flexible enough to allow teachers to adapt content, methods and materials to different local needs, learning environments as well as to their students' needs and interests.

Moreover, curricula should encourage students to engage in continuous learning. Students should be taught not only the knowledge and skills they need today, but also to how to decide what they should continue to learn in the future.

### *Compulsory*

While official policies state that women have the same opportunity as men to study engineering, taking Hong Kong as an example, less than 5% of female students take technology and engineering-related subjects at the secondary level. As mentioned above, although universities do not restrict female students from studying engineering, the constraints of their secondary education put them at a disadvantage in terms of both knowledge and motivation. Moreover, it is insufficient simply to offer extra-curricular activities and interest groups that allow female students to get a taste for engineering. Therefore, it should be compulsory for both male and female students, as well as those with disabilities, to take engineering subjects at the secondary level, so that they can make informed choices about studying engineering at the university level. This approach will realise the objective of the slogan: *engineering for all.*

## Quality Assurance through Assessment Using the Eight Cs

Quality Assurance through assessment is essential. There have been some improvements in the assessment of engineering programmes over the past two decades in Hong Kong, but only through organising assessment activities using the Eight Cs can we assure the quality of engineering programmes in the future.

First, assessment activities should be planned, implemented and evaluated with a view to long-term consequences. Quality Assurance of many engineering programmes in Hong Kong has often been criticised for lacking long-term direction and objectives. This suggests that Quality Assurance is not producing the constructive, long-term suggestions that will improve engineering programmes.

Many educators as well as industry people have criticised the assessment of engineering programmes as being similar to speculation in the stock market. This is reflected in the strange (and gimmicky) titles of many programmes, which are often terminated after only one or two cohorts of enrolling students. This phenomenon recalls the mass media reports that teased the government's recent education slogan, *Education is for Tomorrow* by pointing out that the words imply that education is merely for tomorrow, but not for the day after tomorrow.

Second, it is critically necessary to avoid considering the Eight Cs in a separated or piecemeal way. The *all-round* characteristic is the core spirit and value of Eight Cs. At present, the processes by which we assess the planning, implementation and evaluation of engineering education (whether in terms of overall direction or individual programmes) lack a holistic view and comprehensive consideration. This makes individual engineering programmes unbalanced. As a Chinese saying puts it, *Healing is only for head when headache appears; while healing is only for foot when footache appears.*

Since some programmes focus on promoting students to be competent in a particular area of engineering and technological knowledge, students may lack the opportunity to study and explore in a comprehensive way. In other cases programme developers and teachers may be happy to be successful in providing compulsory engineering programmes or subjects to both male and female students. However, programme content may easily become *normalised* and *averaged.* In addition, many programmes promote neither *particularity*, *creativity* nor *curiosity*. Similarly, encouraging teamwork can promote collaboration, but sometimes it can make students rely too much on others' contributions, so that individual critical thinking is suppressed (Siu 2007). Because of these threats, it is important to have a careful assessment checklist based on the Eight Cs.

## Case Study in Hong Kong

An Eight Cs assessment has been carried out since the programme planning stage at one of the new engineering (and design) programmes jointly organised by a design school and an engineering department of The Hong Kong Polytechnic University. Each year, the Eight Cs criteria that are applied to the implementation of the programme are themselves reviewed. The review asks (a) what, how and whether the criteria have been considered and met; (b) what, how and whether changes are necessary in the programme's internal situation, such as the number of student intake and students' objectives; (c) what, how and whether changes have taken place in external factors, such as the social situation, industry needs and overall educational needs; (d) what has been, and how and whether the relationship has changed among factors (b) and (c) and (e) what and how additional criteria can be related to the Eight Cs, and whether these criteria can be fitted or incorporated into the Eight Cs.

Obviously, Quality Assurance that has a long-term approach and all-round perspective does not imply *inflexible* planning and implementation. Engineering education objectives must be dynamic to respond to continuous social changes; Quality Assurance in engineering education must also be dynamic to cope with continuing change.

The weightings of each of the Eight Cs must vary with the different natures and objectives of individual programmes, and change with the social, industrial and educational changes. This does not imply that curriculum planners and teachers can take *flexibility* as an excuse to eliminate or ignore any of the Eight Cs. Instead, careful thought is required before applying the Eight Cs in Quality Assurance schemes and continuous review of the relative balance of the Eight Cs to fit changing needs is necessary.

## Conclusions

We are living in an ever-changing global community where engineering education has to continuously adapt to social, industrial and educational needs. Curriculum development, assessment and Quality Assurance go hand-in-hand in helping engineering education to successfully adapt to changing circumstances. In fact, the case of Quality Assurance in engineering education presented above is not bounded by the particular situation in Hong Kong. Indeed, the experience can be applied to many places. In order to insure the quality of engineering curricula and to guide continuous improvement, Quality Assurance should be based on curriculum development and assessment related to the *all-round perspective* of the Eight Cs: competent, comprehensive, critical, creative, curious, collaborative, continuous, and compulsory.

Only through improvement in both curriculum development and assessment can engineering education provide the maximum benefit for students, industry and society. And only when improvements are ongoing can we claim that our engineering education is of high quality and able to equip our students to contribute to educational, social and economic development and international competitiveness.

**Acknowledgements** I would like to thank the research funding support of The Hong Kong Polytechnic University and the faculty and researchers of the Massachusetts Institute of Technology, and the National University of Singapore for their comments on the research direction of the projects which generated the findings for this article. The funding from the Asian Scholarship Foundation the K.C. Wong Education Foundation supported the core study in Hong Kong and supplementary studies in other Asian cities. Part of this article is revised from the article published in the 25th Anniversary Commemorative Album of the Hong Kong Examinations and Assessment Authority. I would also like to thank the comments of the examination officers of the Authority.

## References

Adair, J. (1990). *The challenge of innovation.* Surrey: Talbot Adair Press.

Einstein, A. (1938). *The evolution of physics.* New York, NY: Simon & Schuster.

Gilbert, R. (1998). *Living with art* (5th ed.). New York, NY: McGraw-Hill.

Hong Kong Special Administrative Region Government. (1997). *The 1997 Policy Address.* Hong Kong Special Administrative Region Government policy programmes.

Mendelowitz, D. M. (1981). *Children are artists.* Stanford, CA: Stanford University Press.

Parnes, S. J., & Harding, H. F. (Eds.). (1962). *A source book for creative thinking.* New York, NY: Scribner.

Siu, K. W. M. (1997). Criticism and theory studies in design and technology teacher education programmes. *Curriculum Forum, 7*(1), 49–58.

Siu, K. W. M. (2000a). A case study of the difficulties and possibilities for students to initiate their project titles. In K. Volk, W. So & G. Thomas (Eds.), *Science and technology education conference 2000 proceedings* (pp. 112–120). Hong Kong: The Hong Kong Institute of Education.

Siu, K. W. M. (2000b). A study of the relationship between the colour vision deficiency and creativity of design and technology students. *The Korean Journal of Thinking and Problem Solving, 10*(1), 21–30.

Siu, K. W. M. (2000c). The eight percent of design and technology students in our class. *The Journal of Design and Technology Education, 5*(1), 20–25.

Siu, K. W. M. (2001). What should be solved? *The Korean Journal of Thinking and Problem Solving, 11*(2), 9–22.
Siu, K. W. M. (2002). Meeting the new needs: Curriculum development and assessment of technology subjects. In *25th Anniversary Commemorative Album of the Hong Kong Examinations and Assessment Authority* (pp. 48–54). Hong Kong: Hong Kong Examinations and Assessment Authority.
Siu, K. W. M. (2007). Balance in research and practice: Critical reform of research studies in industrial and product design. *Global Journal of Engineering Education, 11*(1), 15–27.
Starko, A. J. (2000). Finding the problem finders: Problem finding and the identification and development of talent. In R. C. Friedman & B. M. Shore (Eds.), *Talents unfolding: Cognition and development* (pp. 233–249). Washington, DC: American Psychological Association.
Wertheimer, M. (1959). *Productive thinking*. London: Tavistock.
Zhu, G., & Shen, S. (Eds.). (2001). *Engineering education in the 21st century* (1st ed.). Beijing: Gao Deng Joao Yu Chu Ban She.

# Engineering Education Quality Assurance: The Essential Pillar of Higher Education Reform in Lithuania

A.V. Valiulis and D. Valiulis

**Abstract** The purpose of this chapter is to present a review of Lithuanian engineering education and its related evaluation and accreditation systems. Lithuanian engineering education has changed dramatically in recent years: initially in 1991 after Lithuania separated from the Soviet Union, then in 1999 when Lithuania accepted the Bologna process and very recently in 2004 upon Lithuania's entry into the European Union. This chapter examines the strengths and weaknesses of the current educational system and, in particular, the assessment of engineering education in the hope that it will enable the further integration of Lithuanian engineering education into the European Higher Education Area. With the 2010 deadline for implementation of the Bologna Process approaching Lithuanian higher education, it is attempting to provide responses to the following questions: What current challenges will persist well into the next few years? Where is European higher education heading? What opportunities is Lithuania facing in an increasingly globalised World?

## Introduction

Society expects a lot from engineers. They are expected to have a strong scientific background; competent technical skills; a sharp awareness of the social concerns linked with their profession roles; a deep appreciation for safety and security; an ethical sense and appropriate behaviour; an openness to other cultures; a willingness to be both geographically and professionally mobile; adequate project management and teamwork skills as well as many other attributes. Existing guidelines for Quality Assurance and assessment in European engineering schools stress the importance of these traits and their documentation in order to increase transparency

A.V. Valiulis (✉)
Vilnius Gediminas Technical University, J. Basanaviciaus Street 28, LT-03224, Vilnius, Lithuania
e-mail: algirdas.valiulis@adm.vgtu.lt

A.S. Patil and P.J. Gray (eds.), *Engineering Education Quality Assurance: A Global Perspective*,
DOI 10.1007/978-1-4419-0555-0_15, © Springer Science+Business Media, LLC 2009

and promote mutual recognition of programmes among engineering education institutions through the European Higher Education Area (EHEA).

The path toward these goals is provided by the European System for Accreditation of Engineering Education (EUR-ACE) project (European System for Accreditation of Engineering Education 2008). As described in the chapter "EUR-ACE: The European Accreditation System of Engineering Education and Its Global Context" by Augusti, EUR-ACE is intended to establish a system for accreditation with the following main purposes: to provide an appropriate European label to the graduates of the accredited educational programmes, to improve the quality of educational programmes in engineering and to facilitate their trans-national recognition. This system for accreditation will be facilitated by European Union (EU) directives and will be a significant contribution to the harmonisation of European higher engineering education as set forth in the Bologna process. Quality Assurance, assessment and accreditation of Lithuanian higher engineering education take place within the context of the EUR-ACE.

## Lithuanian System of Higher Engineering Education

Higher education (including engineering education) occupies the highest stage in the system of consecutive education in Lithuania. The Lithuanian higher education system is comprised of both university and colleges that are either state or privately supported. Universities offer Bachelors, Masters and Doctoral programmes, while colleges offer one-level undergraduate studies.

Study programmes are registered by the Ministry of Education and Science. The quality of programmes and the quality of research is periodically evaluated by a national institution, namely, the Centre for Quality Assessment in Higher Education (CQAHE). The mutually supportive areas of engineering studies and fundamental research constitute the foundation of higher engineering education.

Lithuania has had a three-cycle model since 1993 when a fundamental reform was introduced into higher education, including engineering. The previous model was one of the continuous higher educations over a period of 5 years. A new two-cycle structure, based on an Anglo-Saxon model was introduced in 1993, with English nomenclature for qualifications: Bachelors, Masters and Doctorate (Minister of Science and Education 2005). In essence, the 5-year model was replaced by a 4+2 structure, except for first-cycle Bachelors degrees in colleges. This 4+2 model of Bachelors and Masters degrees is capped by a doctoral cycle, lasting for 4 years.[1]

[1] Law on Higher Education of the Republic of Lithuania (Valstyb.zinios, 2000, No. 27-715, Valstyb.zinios, 2001, No. 16-496, Valstyb.zinios, 2002, No. 3-75, Valstyb.zinios, 2002, No. 71-2968, Valstyb.zinios, 2003, No. 47-2058, Valstyb.zinios, 2005, No. 85-3136).

Studies are measured in credits based on the European Credit Transfer and Accumulation System (ECTS),[2] with 1 credit equal to 40 related hours of student work (contact hours, laboratory work, independent preparation, etc.). The average volume of 1-year full-time study amounts to 40 credits. An undergraduate engineering study programme at the university level in Lithuania comprises 140–180 credit points (210–270 ECTS). Masters studies comprise 60–80 credit points (90–120 ECTS). The non-university or college level study programmes consist of at least 120–160 credits (180–240 ECTS). And, as noted, the duration of doctoral studies should not exceed 4 years.

The Lithuanian credit system is fully compatible with ECTS. The question is whether it is fully compatible in respect to credit accumulation. The norms for student workload across the EHEA, as reported by the Tuning project (Gonzélez and Wagenaar 2003a, b), are that 1 credit is equivalent to a minimum of 25 h and a maximum of 30 h. In a general sense, it is clear that 1 Lithuanian credit is equal to 1.5 ECTS credits. A student in Lithuania studies for a total of 1,600 h in an academic year of normal length, a length which is compatible with the norms suggested in the ECTS Users Guide, where 38–40 weeks are given as the EHEA average. Therefore, in ECTS terms, 1 Lithuanian credit is equal to 26.66 h of work.

The Bologna Accord documents emphasise five key issues that are to be addressed by higher education institutions: recognition, accessibility, conduct of scientific research, attractiveness and openness to society and flexibility (as opposed to regulation). During the 17 years since restored independence in 1991, Lithuania has accumulated considerable experience in developing its national higher education system consistent with the Bologna process. The active involvement of the academic community in the establishment of the EHEA fostered many changes in Lithuanian higher engineering education which have been recognised at the Bergen (2005) and London (Bologna Process 2007) summits of European higher education ministers. In 2007, the ministers rated Lithuania higher education as good or very good in all three priority spheres, namely (1) preparing students for life as active citizens in a democratic society; (2) preparing students for their future careers and enabling their personal development and (3) creating and maintaining a broad, advanced knowledge base by stimulating research and innovation.

The Development Plan of the Lithuanian Higher Education System for the Period from 2006 to 2010 states the following (Government of the Republic of Lithuania 2006):

> …striving to implement the provisions of strategic EU and Bologna process documents, in order to be competitive in the European and world knowledge markets, it is essential to clearly perceive problems and the inevitability of changes, to possess a future orientation, and to have definite strategies and solutions at both the state and institutional levels. It is necessary to pay more attention to the analysis of changes in the higher education system, not only by evaluating achievements but also by identifying problems and their consequences.

---

[2]European Credit Transfer and Accumulation System (ECTS) is a standard for comparing the study attainment and performance of students of higher education across the European Union and other collaborating European countries. See http://ec.europa.eu/education/index_en.htm

## The Main Weaknesses, Threats and Opportunities of Lithuanian Higher Engineering Education

There are several weaknesses evident in Lithuanian higher engineering education. For example, higher education institutions have established engineering study programmes where the curriculum has not been substantiated by sound research and is not related to the needs of the market place or the development of the society. Problems with curricula are compounded by the facts that the average age of those currently in the academic community is increasing and there are few incentives for young people to pursue careers in academe; the gap between higher education institutions and the business sector is widening; there is a lack of skills and experience in innovation, technology transfer and commercialization and there is a shortage of modern laboratory equipment for study and research.

There are also weaknesses related to financial resources (Government of the Republic of Lithuania 2004, 2006). The 2007 European average financial allocation per student was seven times larger than that in Lithuania. In addition, mechanisms for the internal management of institutions are not flexible; in that state funding is earmarked for precisely defined purposes, which limits a university's freedom to allocate its resources in the most appropriate way. In addition, being part of the EU has meant increasing competition for students from other higher education institutions.

Other threats include low salaries and poor work environments in Lithuania in comparison with other EU countries which have resulted in the migration of teachers and students causing a *brain drain*. At the same time, there has been a decline of interest in engineering education programmes among secondary school leavers, and demographic trends point toward a smaller group of students entering higher education in the near future.

EU integration presents opportunities as well as threats for Lithuanian higher education institutions. Accession to the EU and the Bologna process opens new opportunities for partnership with other universities in both research and studies. Access has been gained to the financial resources of the EU structural and other funds. Rapid economic growth, at least up to 2008, and strengthened cooperation with internal and external business companies have also resulted in opportunities to provide new and expand existing engineering studies, research and in-service training. It is within this context that the development of Quality Assurance systems in Lithuania has taken place.

## The General Context for Lithuanian Higher Education Quality Assurance

Lithuanian higher education legislation forms the general context for Quality Assurance at the national level. The review of Lithuanian legislation by foreign experts has been conducted from three different but interrelated points of view (1) the coherence of current Lithuanian legislation on higher education institutions and study programmes, (2) the degree to which Lithuanian legislation conforms to the exigencies of the Bologna process and (3) comparisons, where appropriate, with the legislation of other countries in the EHEA. The findings indicate that the Law on Higher Education defines the national system of higher education quite clearly. However, there is no attention paid to the Bologna Process and Lisbon Strategy (Norwegian Ministry of Education and Research 2005).

As noted above, the Lithuanian credit system is fully compatible with ECTS. However, the ECTS norms for hours worked have not been incorporated into Lithuanian legislation. The question is whether it is necessary to do so, since very few countries have taken this step. By and large, the norms remain primarily a concept advanced in the ECTS Users Guide, as recommended by Tuning project.

The Dublin Descriptors, which define the higher education cycles by specifying the levels of competence to be achieved, have not been incorporated into legislation either. Incorporating the Dublin Descriptors is an essential step in establishing the three-cycle pattern of higher education (Joint Quality Initiative 2004). The Dublin Descriptors are benchmarks, i.e. the points of reference for all other types of descriptors (Qualification Descriptors, Programme Descriptors. Level Descriptors and Module Descriptors). These descriptors may be used to set benchmarks in areas such as engineering education. However, since the Dublin Descriptors were only formally approved in May 2005, one would not expect them to have been to incorporate into legislation as of yet. Even so, the Dublin Descriptors should be written into Lithuanian legislation as soon as possible.

While formal legislation provides the national context for Quality Assurance, study programme curricula provide the local, institutional context. A well-constructed curriculum facilitates the process of Quality Assurance.

The traditional way of designing study modules and programmes was to start with course content. Teachers decided on the content that they intended to teach and then assessed how well the students had learned the material. Modern trends in higher education represent a shift from the traditional teacher-centred approach to a student-centred, learning-centred or outcomes-based approach that focuses on what the student are expected to be able to do at the end of the module or programme (Gosling and Moon 2001).

The Dublin Descriptors are statements of learning outcomes while ECTS describe the credits related to studies in terms of student workload. Together they provide a foundation for curriculum development that is compatible with the Bologna Process. With the implementation of the Bologna Process by 2010, all modules and programmes throughout participating Bologna process countries, including Lithuania will need to be described using the outcomes-based approach, which provides the foundation for assessment and Quality Assurance.

## The Assessment of Higher Engineering Education Quality

The Law on Higher Education and Research stipulates the main requirements for Lithuanian higher education institutions and defines the elements of Quality Assurance in engineering education. The Law identifies internal and external structures intended to assure quality.

The main internal Quality Assurance structures include the Senate, which is responsible for the management and control of the quality of studies and research, and the University Council, which evaluates higher education institutions' goals and their implementation, as well as the contribution that institutions make to the economic, social and cultural growth of the country. Institutional assessment in Lithuania will begin in a few years. The Institutional Evaluation Programme (IEP) launched by the European University Association (EUA) has been selected by many Lithuania higher education institutions as the most appropriate means of Quality Assurance (European System for Accreditation of Engineering Education 2008).

Engineering study programmes are assessed in two ways:

- New programmes are assessed internally according to the quality of the registration proposal documents
- Registered programmes are assessed based on a self-study report and the results of a visit by a panel of external experts

External assessment is performed by evaluating the quality of studies and research quality separately. The main Lithuanian governmental institutions responsible for organising the external assessment of higher education study programme quality are the Ministry of Education and Science and the CQAHE.[3]

A programme self-study must address the programme goals and tasks, material conditions, internal process for Quality Assurance, external relations and feedback mechanisms. A group of external experts reviews the self-study report and visits the programme. External experts may be selected from educational and science institutions, ministries and other state institutions and offices, professional associations and creative organisations, as is appropriate.

Having reviewed the programme self-study and conducted a visit to the institution, the group of experts prepares an assessment report. The conclusions are presented to the institution and the study programme. There is an opportunity for the institution to draw the experts' attention to any factual mistakes in the report. After receiving the institution's response to the assessment, the group of experts prepares a final report, which is presented to the Centre of Quality Assessment. The conclusions are discussed at the Board of Experts for the Assessment of Activities of Science and Study Institutions. Subsequently, the final conclusions and the decision of the Board of Experts are sent to the Ministry of Education and Science. Based on these documents, the Ministry issues a decree on the certification of the assessed study programme. The final assessment conclusions are made public in the information

[3]Centre for Quality Assessment in Higher Education (CQAHE) is an independent public agency established in 1995. The Centre implements the external Quality Assurance policy in research and higher education in Lithuania and contributes to the development of human resources by creation of enabling conditions for free movement of persons. The Centre was founded by the Ministry of Education and Science of the Republic of Lithuania as an expert institution. Centre for Quality Assessment in Higher Education (CQAHE): http://www.skvc.lt/en/default.asp?id=0. Accessed December 2008.

publications of the Centre for Quality Assessment (Government of the Republic of Lithuania 2004b).

Study programmes are certified *unconditionally*, *conditionally* or *restricted*. The programmes that are certified *unconditionally* are free to function until the next certification. Those certified *conditionally* have to be amended within a given period and then presented again for certification. If the activity of an institution department, study programme, etc. is found to be of especially low quality, it may be *restricted* or *terminated*.

The external assessment of studies programmes started in academic year 1999–2000. At present 79% programmes are certified *unconditionally*, 17% are certified *conditionally* and 4% are *restricted* or *not certified*. External assessment is an essential aspect of the Quality Assurance process, since it both makes a judgement of the quality of study programmes and points out areas in need of improvement.

Based on the results of this exhaustive external assessment process, the CQAHE provides recommendations for the Ministry of Education and Science about the certification of study programmes. In effect, the results of the external assessment of study programmes are de facto assessments of the capacity of the university to perform studies in particular fields. However, at present such assessment does not fully reflect the capacity of universities to conduct studies in specific science areas.

## Conclusions

European Union policy in the field of higher education emphasise the establishment of the Europe Higher Education Area (EHEA). The processes of EU higher education integration will have a significant impact on Lithuania's higher engineering education and research programmes. One of the most important of which is the expectation that Quality Assurance should focus not only on structural issues but also on learning outcomes. Such activities are as of yet underdeveloped and are not yet properly regulated by Lithuanian legislation.

In addition, despite of noteworthy advances, results to date show that Lithuania's accreditation system is not well developed. Further efforts are therefore necessary to improve the accreditation system in Lithuania so as to reconcile it with the international Quality Assurance procedures approved by the European Commission.

## References

Bologna Process. (2007). *London Communiqué, towards the European higher education area: Responding to challenges in a globalised world, 18 May 2007*. Accessed December 2008. http://www.cicic.ca/docs/bologna/2007LondonCommunique.en.pdf

European System for Accreditation of Engineering Education. (2008). *Implementation of a European system for accreditation of engineering education with the contribution of the*

*European commission DG education and culture*. 01.09.06-31.07.08. Accessed December 2008. http://www.feani.org/EUR_ACE/EUR-ACE%202/EUR_ACE2_Main_Page.htm

Gonzélez, J., & Wagenaar, R. (Eds.). (2003a). *Tuning educational structures in Europe: Final report phase one*. University of Deusto and University of Groningen. Accessed December 2008. http://www.unist.hr/ects/povjerenstvo/TuningECTS.pdf

Gonzélez, J., & Wagenaar, R. (2003b). *Tuning educational structures in Europe: A pilot project supported by the European Commission in the framework of the Scorates programme*. Accessed December 2008. http://eua.uni-graz.at/Tuning_summary_EUA_conference.pdf

Gosling, D., & Moon, J. (2001). *How to use learning outcomes & assessment criteria*. London: SEEC.

Government of the Republic of Lithuania. (2004). *Methodology for establishing the need for financing from the budget of the Republic of Lithuania and for allocation of funds for academic and study institutions*. Valstyb.zinios, 2004, No. 151–5493.

Government of the Republic of Lithuania. (2006). *On the adoption of the development plan for 2006–2010 of Lithuanian higher education system*. Adopted by the resolution No. 335 of the Government of the Republic of Lithuania of 5 April 2006, Valstyb.zinios, 2006, No. 39–394.

Joint Quality Initiative. (2004). Shared Dublin descriptors for Short Cycle, First Cycle, Second Cycle and Third Cycle Awards (2004), Draft 1. *Working document on JQI meeting in Dublin on 18 October 2004*. Accessed December 2008. http://www.google.lt/search?hl=lt&q=Shared+Dublin+descriptors+for+Short+Cycle%2C+First+Cycle%2C+Second+Cycle+and+Third+Cycle+Awards&btnG=Paie%C5%A1ka&meta=

Minister of Science and Education. (2005). *Description of General Masters Programmes' requirements*. Approved by Order No ISAK-1551 by the Minister of Science and Education, 22 July 2005.

Norwegian Ministry of Education and Research. (2005). *Report from a working group appointed by the Bologna follow-up group to the conference of European Ministers responsible for higher education, Bergen, 19–20 May 2005*. Accessed December 2008. http://www.bologna-bergen2005.no/Bergen/050509_Stocktaking.pdf

# General Approaches and Techniques

# Using a Measure of Student Holistic Development for Quality Assurance

**Larry A. Braskamp**

**Abstract** Engineering is a global profession and engineering education has been charged to prepare engineers accordingly. Students' knowledge and acquisition of technical skills in the field of engineering are necessary but insufficient in the preparation of future engineers. Personal and interpersonal skills and attitudes are also worthy and essential requirements in the preparation of engineers. Professional and other accrediting and Quality Assurance agencies which have the responsibility for the Quality Assurance of engineering programs are asking institutions to be held accountable for meeting these expanded and inclusive requirements. However, dimensions and domains of student and human development are some of the more difficult to be assessed, and are thus not well represented in the arsenal of instruments and strategies that focus on student development, more broadly defined. This chapter describes the work done in constructing an instrument, the Global Perspective Inventory that assesses students' progress in their journey in becoming global citizens.

## Introduction

C. Judson King, a chemical engineer and former provost and senior vice president for academic affairs of the University of California system, in his article, *Let Engineers Go to College*, begins with these words (King 2006 p. 1),

> The challenges that engineers will face in the 21st century will require them to broaden their outlooks, have more flexible career options, and work closely and effectively with people of quite different backgrounds. ... The... nature of engineering careers in the United States are changing in fundamental ways. The issues with which engineers engage have become more and more multidimensional, interacting with public policy and public perceptions, business and legal complexities, and government policies and regulations, among other arenas. This is the natural result of technology becoming more and more pervasive in society and politics.

L.A. Braskamp (✉)
Loyola University Chicago, 820 N Michigan Avenue, Chicago, IL 60611, USA
e-mail: braskamp@yahoo.com

A.S. Patil and P.J. Gray (eds.), *Engineering Education Quality Assurance: A Global Perspective*,
DOI 10.1007/978-1-4419-0555-0_16, 

King is not alone in this perspective about the preparation of future engineers, however. The engineering profession in the United States, as reflected by its accreditation council, Accreditation Board for Engineering and Technology (ABET), includes these statements in its criteria of excellence for postsecondary educational programs that desire to be accredited by ABET: Graduates should have the broad education necessary to understand the impact of engineering solutions in a global social context and knowledge of contemporary issues (ABET 2000). Moreover, the recognition of a broad-based education of future engineers has a world-wide endorsement. For example, the Engineering Council of South Africa (ECSA) has this requirement in its standards for accrediting programs that prepare students for a bachelor degree in engineering; "A graduate must be critically aware of the impact of engineering activity on society and the environment" (ECSA 1998).

Because engineering is a global profession, engineering education has been charged to prepare engineers accordingly. However, it is widely recognized that engineering knowledge and the acquisition of technical skills are necessary but insufficient in the preparation of future engineers. Of equal importance are personal and interpersonal skills and attitudes.

## Challenges to these Expanded Expectations and Requirements

Throughout the history of accreditation, cognitive knowledge and the acquisition of technical skills in science, technology, engineering, and mathematics (STEM) have often been assessed. This is due to the fact that engineering educators and professional practitioners are generally in agreement on the necessary knowledge and technical skills needed to be an engineer and on the measurement of these qualities. Developmental behaviors and attributes such as personal and interpersonal skills and attitudes are both more difficult to agree on and to assess. For example, values such as honesty, ethical reasoning, commitment to the public good, civility, and hard work are all reasonable virtues and regarded as worthy goals in an education of a further engineer, but to date there is little research and few instruments and measurement strategies available to assess their attainment. In short, dimensions and domains of personal and interpersonal skills and attitudes are some of the more difficult to be assessed, and are thus not well represented in the arsenal of instruments and strategies for assessing student development.

This chapter then focuses on the work done in constructing an instrument to assess students' progress in becoming global citizens, an important element of both personal and professional development within the context of today's global engineering community. The instrument in question is the Global Perspective Inventory (GPI). In essence, the creation of the GPI was based on the developmental perspective that students are on a journey during college. In this journey, students are given opportunities to reflect on three big questions:

- How do I know?
- Who am I?
- How do I want to relate with others?

*How do I know?* reflects the cognitive dimension; a dimension that represents thinking, knowing, the mind, and the *head*. This dimension of development is centered on one's knowledge and understanding of what is true and important to know. It includes viewing knowledge and knowing with greater complexity; no longer relying on external authorities to have absolute truth; moving from absolute certainty to relativism when making judgments; and commitments within the context of uncertainty.

*Who am I?* reflects the intrapersonal dimension, often referred to the spirit, being sense of self, and the *heart*. This dimension of development focuses on one becoming more aware of one's personal values and integrating them into one's self-identity. This journey leads to a sense of self-direction and purpose in life; becoming more aware of one's strengths, values, and personal characteristics; and having a sense of self, and viewing one's development in terms of this self-identity.

*How do I relate to others?* reflects the interpersonal dimension, a dimension that focuses on the social context, the behavioral, and the *hands*. This dimension of development is centered on the willingness to interact with other people who have different social norms and cultural backgrounds, an acceptance of others, and a feeling of being comfortable when relating to others. It includes being able to view others as unique and seeing one's own uniqueness. And finally, this dimension embodies a way of relating to others that progresses from dependency to independence to interdependence.

## Description of Global Perspective Inventory Scales

There are six scales in the GPI, two for each of the three domains: Cognitive, Intrapersonal, and Interpersonal. Selected items of each of the scales are shown later in this section. (The online version can be seen by going to http://gpi.central.edu, clicking on *Complete it* and use the access code, 9910.)

The *cognitive domain* consists of *knowing* and *knowledge*. The knowing scale measures the complexity of one's view and the importance of cultural context in judging what is important to know and value. The knowledge scale measures degree of understanding and awareness of various cultures as well as an understanding of these cultures' impact on our global society and language proficiency in a second language.

The two scales of the intrapersonal domain are *identity* and *affect*. The identity scale assesses an awareness of an individual's uniqueness and the degree of acceptance of one's ethnic and racial backgrounds as well as one's gender and lifestyle. The affect scale measures the level of a person's self-confidence and respect and acceptance of cultural perspectives different from his or her own.

The interpersonal domain scales are *social interactions* and *social responsibility*. The social interactions scale examines a person's degree of engagement with others who are different and the degree of cultural sensitivity in living in a pluralistic setting. The social responsibility scale examines the level of commitment to the common good and the degree of interdependence with others.

Specific items of six scales are as follows:

1. Cognitive: Knowing
    - When I notice cultural differences, my culture tends to have the better approach (reverse).
    - In different settings what is right and wrong is simple to determine (reverse).
    - I can evaluate issues from several different perspectives.
    - I prefer complex rather than straightforward interpretations of debatable issues.
2. Cognitive: Knowledge
    - I am informed of current issues that impact international relations.
    - I understand the reasons and causes of conflict among nations of different cultures.
    - I can discuss cultural differences from an informed perspective.
    - I am aware of how other cultures consider "fairness" differently from my own culture.
3. Intrapersonal: Identity
    - I have a definite purpose in my life.
    - I know who I am as a person.
    - I see myself as a global citizen.
    - I do not feel threatened emotionally when presented with multiple perspectives.
4. Intrapersonal: Affect
    - I can explain my personal values to people who are different from me.
    - I am confident that I can take care of myself in a completely new situation.
    - I prefer to work with people who have different cultural values from me.
    - I am accepting of people with different religious and spiritual traditions.
5. Interpersonal: Social Interaction
    - Most of my friends are from my own ethnic background (reverse).
    - People from other cultures tell me that I am successful at navigating their cultures.
    - I enjoy when my friends from other cultures teach me about our cultural differences.
    - I am open to people who strive to live lives very different from my own life style.
6. Interpersonal: Social Responsibility
    - I think of my life in terms of giving back to society.
    - I work for the rights of others.
    - I put the needs of others above my own personal wants.
    - I consciously behave in terms of making a difference.

In sum, students (and all human beings) are simultaneously thinking and acquiring more knowledge, seeking a sense of self, and interacting with others as they try to make sense of and give meaning to their journey of life. They are using *their heads*, *their hearts*, and *their hands* in their journey toward becoming a unique whole person.

## Interpreting the Results of the GPI

The first major principle that should be stressed in the use of the GPI as an assessment instrument is that assessment is a service to users. In its best use, assessment promotes discussion, rather than decisions. Thus the use of the GPI is intended to be a staring point for discussion among the major stakeholders – faculty, administrators, practitioners, and students. Those closest to the action of promoting students' learning and development need to be highly involved in the assessment, especially in the interpretation of the results. Ownership of the process is critical for maximal use by the intended users.

This suggests that assessment evidence needs to be collected and presented in ways that facilitate and promote discussion about the goals of a program and the success of students engage in related interventions. To enhance the utilization of the GPI in this regard, a strategy has been adopted to help users better understand the idea of developing global citizens and to plan how students are to become global citizens. This strategy has these features.

First, a report has been designed to provide frequency distributions of all of the items, average scores, and standard deviations (an index of the variability of the responses to the scale scores) for each group of students. For example, the results from entering freshmen students can be compared with graduating seniors at a school of engineering and with students with other majors at the same institution. (A sample Interpretative Guide and Sample Report can be viewed by accessing http://gpi.central.edu and clicking on *Resources* and then *Guide*.) Since the GPI is a survey of 46 fixed response or forced-choice items classified into six scales, it economically obtains evidence about students' development. However, although the GPI is theory based, GPI provides indirect evidence of students' growth and development. Therefore, corroborating evidence is required.

The second part of the strategy is an Interpretative Guide that accompanies the results. The guide focuses on the connections between campus and program environmental conditions and three dimensions of development – cognitive, intrapersonal, and interpersonal – measured in the GPI. (This connection is described in the next section.) Together with the report of results, the guide is intended to promote discussion of the evidence among the key stakeholders by stressing that the context of the evidence is critical in understanding the results. That is, it emphasizes the idea of *sitting beside* (i.e., discussion) rather than *standing over* (i.e., decisions), which is the main theme in our work on assessing faculty as well as students and programs (Braskamp and Ory 1994).

Third, we encourage users of the results to tell stories about why they think the results are what they are. In essence, we encourage users to treat evaluation and assessment as telling stories with evidence. The primary users of the GPI results will benefit by asking questions about the actual and hoped for connections between what students report on the three dimensions of their development – thinking, being, and relating – and what they know about the campus and local program environment. That is, decision makers are encouraged to view the GPI results, as only one source of information is useful in guiding changes in a program's sociocultural environment in order to achieve the students' learning and development goals they desire. Thus we encourage the users to develop stories by adding other evidence to inform their understanding of the results.

The following questions are to help users of a GPI report focus on the way an environment can optimally influence students so they will more readily meet the expectations of an engineering program:

1. Cognitive: How Do I know?

   - How can you help students better understand how their own cultural backgrounds makes assumptions about the role of authority and what is good and truthful?
   - How can you better assist students compare their personal values, practices, and behaviors, norms and expectations with those of other countries and nations?
   - How can you better encourage students to reflect on the issue that people from different cultures and countries may think differently about the role of government, business, religion, family values, schooling, and work?
   - How can you help students see the value of having them exposed to multiple perspectives on an issue or topic and the use of a technical innovation in different nations and cultures?

2. Intrapersonal: Who Am I?

   - How do you help students develop more complex views of themselves, taking into consideration their own cultural backgrounds? Do you give them opportunities to share with others in class and out of class their uniqueness as human beings?
   - How do you encourage students to develop a sense a self that incorporates their own cultural backgrounds and family influences? Do you help them value their pride in their uniqueness?
   - How do you provide opportunities in classes or arrange sessions for students to talk about their own values, sense of self and purpose of life, and relationships with others not like them?

3. Interpersonal: How do I Relate to Others?

   - How do you assist students to be more comfortable in interactions with other students, staff, faculty, and citizens from different cultural backgrounds, values, and points of view?
   - How do you inform and demonstrate to students studying abroad different cultural traditions, practices, and social interactions, especially those that are different from theirs?

Given the holistic view of student development, we encourage the reader to discuss in a holistic way how students progress in their thinking, search for self-identity, and relate to others, taking into account the global learning and developmental goals they desire for their students.

## A Framework for Fostering *Global Citizenship* as an Integral Part of Holistic Student Learning and Development

Of course, it is not enough to just document the status of student's learning and development. Most Quality Assurance schemes require a focus on continuous improvement, based on evaluation results that may involve changes in the structure and content of engineering education programs. Four dimensions form the proposed framework for fostering *global citizenship* that is an integral part of holistic student learning and development (Braskamp et al. 2006). Each of these dimensions represents an aspect of a campus environment: culture, curriculum, co-curriculum, and community within and beyond the campus

*Culture* focuses on the identity and character of the program or campus. It includes the rituals that reflect the traditions and legacies, habits of staff and faculty in their interactions with students, rules and regulations, and physical setting. *Curriculum* focuses on the courses and methods employed in a program by instructional staff. It includes course content (what is taught) and pedagogy that reflects local style of teaching and interactions with students (how content is taught). *Co-curriculum* focuses on the activities out of the classroom that foster student development. It includes planned interventions, programs, and activities such as organized trips, social and cultural events, residence hall living arrangements, internships and practica, emersion experiences, and leadership programs. *Community within campus* focuses on the relationships among the various constituencies including students, faculty, and staff intended to create a sense of camaraderie and collegiality, whereas *community beyond* focuses on the relationships that colleges (or study abroad programs) have with external and local community agencies, schools, and churches.

The chart in Fig. 1 can be used to organize the total student learning and development sociocultural environment and connect it to desired student dimensions of student learning and development. It represents a template to simultaneously select student learning and developmental goals (desired ends) and organize appropriate means.

| | | MEANS | | | |
|---|---|---|---|---|---|
| | | **Culture** | **Curriculum** | **Co-curriculum** | **Community** |
| **ENDS** | Cognitive | | | | |
| | Intrapersonal | | | | |
| | Interpersonal | | | | |

**Fig. 1** Matrix for selecting student learning and developmental goals (desired ends) and organizing appropriate means [Braskamp et al. (2006), Reproduced here with the kind permission of Wiley, Inc.]

## Conclusions

The GPI can provide important evidence to engineering education stakeholders who are responsible for planning and implementing learning environments and for demonstrating the quality of those environments. The GPI measures three interconnected domains of learning and development – cognitive, intrapersonal, and interpersonal. It goes beyond the acquisition of technical engineering knowledge and includes indicators of the attitudes, values, and competences that student need to function and lead in a global profession. These are important elements in many Quality Assurance schemes as noted in other chapters of this book (see for example, the chapters "Taxonomies of Engineering Competencies and Quality Assurance in Engineering Education" by Woollacott and "CDIO and Quality Assurance: Using the Standards for Continuous Programme Improvement" by Brodeur and Crawley).

The use of the GPI is based on a strategy that deliberately connects the three domains of holistic student learning and development with four major components of a campus and program environment – culture, curriculum, co-curriculum, and community. The GPI provides evidence, albeit indirect, about each item and the six scales that can be used by stakeholders to focus discussion on how well they are creating the environment that optimally fosters and enhances the desired student's learning and development. In short, the GPI provides important evidence for Quality Assurance that can be integrated with other types of evidence such as qualitative summaries of interviews and observations as well as direct measures of student learning and development.

We need to better design learning environments that integrate learning and development, i.e., we, our students included, need to think and act from a holistic and integrative perspective. Just being engaged is insufficient and results in an incomplete education. Focused reflection, analysis, and synthesis based on experience are essential for growth (Braskamp 2008). Quality Assurance is not just a one-time judgment. It is essentially a process for determining if a program has the capacity for continuous improvement in order to meet changing demands. This is no where as important as in the field of engineering.

## References

Accreditation Board for Engineering and Technology (ABET) Engineering Criteria. (2000). Accessed 22 August 2008. http://www.abet.org

Braskamp, L. A. (2007). Three "Central" questions worth asking. *Journal of College and Character, IX*(1). Accessed 22 August 2008. http://www.collegevalues.org/pdfs/Braskampthree centralquestions.pdf

Braskamp, L. A. (2008). Developing global citizens. *Journal of College & Character, X*(1).

Braskamp, L. A., & Braskamp, D. C. (2007). Fostering holistic student learning and development of college students: A strategic way to think about it. *The Department Chair, 18*, 1–3.

Braskamp, L. A., & Ory, J. C. (1994). *Assessing faculty work*. San Francisco, CA: Jossey-Bass.

Braskamp, L. A., Braskamp, D. C., & Merrill, K. (2008). *Global perspective inventory*. Accessed 22 August 2008. http://gpi.central.edu

Braskamp, L. A., Trautvetter, L. C., & Ward, K. (2006). *Putting students first: How colleges develop students purposively*. San Francisco, CA: Jossey-Bass.

Engineering Council of South Africa (ECSA). (1998). *Standards for accrediting University Engineering Bachelors Degrees, document PE-61*. Available as a pdf at: http://www.ecsa.co.za/documents/PE-61-r2.pdf

King, C. J. (2006). Let Engineers go to college. *Issues in Science and Technology, 22*(4), 25–28. Accessed 22 August 2008. http://www.issues.org/ss.4/p_king.html

# CDIO and Quality Assurance: Using the Standards for Continuous Program Improvement

**Doris R. Brodeur and Edward F. Crawley**

**Abstract** The CDIO Initiative is a world-wide collaboration of engineering programs at universities in more than 16 countries in the Americas, Europe, Africa, Asia, and Australia. Collaborators have developed a set of 12 standards, or best practices, that characterize the CDIO approach to engineering education and provide the basis for program evaluation. This standards-based program evaluation extends the evaluative criteria of ABET's EC2000 and other outcomes-based approaches. Evidence of overall program value is collected from multiple sources, using both quantitative and qualitative methods. Evidence and results, forming the basis of decisions about the program and its plans for continuous improvement, are important components of most Quality Assurance schemes.

## Introduction

In October 2000, the Massachusetts Institute of Technology, Chalmers University of Technology, the Royal Institute of Technology, and Linköping University launched an initiative to reform engineering education (Brodeur et al. 2002).[1] Sponsored in part by funding from the Knut and Alice Wallenberg Foundation, this initiative has grown to include engineering education programs in more than 16 countries on five continents.

The vision of the CDIO approach is to provide students with an education that stresses engineering fundamentals set in the context of Conceiving, Designing, Implementing, and Operating (CDIO) real-world product, processes, and systems.

D.R. Brodeur (✉)
Department of Aeronautics and Astronautics, Massachusetts Institute of Technology, 77 Massachusetts Avenue, Cambridge, MA 02139, USA
e-mail: dbrodeur@MIT.EDU

[1] This chapter is based on a paper that was copyrighted by the American Society for Engineering Education in the Proceedings of the 2005 ASEE Annual Conference and Exposition.

A.S. Patil and P.J. Gray (eds.), *Engineering Education Quality Assurance: A Global Perspective*,
DOI 10.1007/978-1-4419-0555-0_17, © Springer Science+Business Media, LLC 2009

This context is a generalized description of a complete system lifecycle, called in this approach, Conceive–Design–Implement–Operate. The Conceive stage includes defining the need and technology, considering the enterprise strategy and regulations, and developing the concept, architecture, and business case. The second stage, Design, focuses on creating the design, i.e., the plans, drawings, and algorithms that describe what will be implemented. Implement refers to the transformation of the design into the product, process, or system, including manufacturing, coding, testing, and validation. The final stage, Operate, uses the implemented product, process, or system to deliver the intended value, including maintaining, evolving, and retiring the product, process, or system.

The CDIO Initiative focuses on the reform of curriculum, teaching and learning methods, learning assessment, design–implement experiences, and the creation and retasking of laboratories and workspaces. Two of its major accomplishments include the development of a detailed and validated list of learning specifications, called the CDIO Syllabus, and a set of standards, or best practices, that characterize the essential features of the CDIO approach. Descriptions of the CDIO approach to engineering education and its global implementation can be found at http://www.cdio.org. After a brief description of the CDIO Syllabus, this chapter focuses on the use of the CDIO Standards for program evaluation, Quality Assurance, and continuous improvement.

## *Student Learning Outcomes Assessment and Standards-Based Program Evaluation*

In the educational evaluation literature, program evaluation is sometimes referred to as program assessment. However, the CDIO Initiative uses the term evaluation to mean a judgment of the overall value of a program based on evidence of a program's progress toward attaining its goals. We apply the term assessment to the measure of the extent to which each student achieves specified learning outcomes. Instructors usually conduct this assessment within their respective courses.

We recognize that the terms are sometimes used interchangeably, but for the sake of clarity, we use evaluation in relation to programs, and assessment for student learning at the course level (see chapter "Quality Assurance in European Engineering Education: Present and Future Challenges" by Cowan, for a similar definition of these terms). Many discipline accreditation groups, e.g., ABET, as well as the various US regional accrediting organizations, have moved in the direction of outcomes-based program and institutional evaluation as a means of Quality Assurance.

The CDIO Syllabus provides the foundation for the assessment of student learning and, ultimately, the evaluation of a program's value or worth (i.e., quality) in terms of its success in fostering desired student learning outcomes. The Syllabus is discussed next; followed by a description of the CDIO Standards that provide the basis for program evaluation.

## *The CDIO Syllabus*

The CDIO Syllabus is a detailed list of knowledge, skills, and attitudes rationalized against the norms of contemporary engineering practice, comprehensive of all known skills lists, and reviewed by experts in many fields. The principal value of the Syllabus is that it can be applied across a variety of programs and can serve as a model for all programs to derive specific learning outcomes. As is shown in the next section, CDIO Standard 2 emphasizes the importance of the Syllabus in engineering education reform.

Standard 2: Learning Outcomes

Specific, detailed learning outcomes for personal and interpersonal skills, and product, process, and system building skills, as well as disciplinary knowledge, consistent with program goals and validated by program stakeholders

The knowledge, skills, and attitudes intended as a result of engineering education, i.e., the learning outcomes, are codified in the CDIO Syllabus (Crawley et al. 2007). These learning outcomes detail what students should know and be able to do at the conclusion of their engineering programs. As shown below in the CDIO Syllabus, in addition to learning outcomes for technical disciplinary knowledge (Section 1), the Syllabus specifies learning outcomes as personal and interpersonal; and product, process, and system building. Personal learning outcomes (Section 2) focus on individual students' cognitive and affective development, which include engineering reasoning and problem solving, experimentation and knowledge discovery, system thinking, creative thinking, critical thinking, and professional ethics. Interpersonal learning outcomes (Section 3) focus on individual and group interactions such as teamwork, leadership, and communication. Product, process, and system building skills (Section 4) focus on conceiving, designing, implementing, and operating products, processes, and systems in enterprise, business, and societal contexts.

Learning outcomes are reviewed and validated by key stakeholders – groups who share an interest in the graduates of engineering programs – for consistency with program goals and relevance to engineering practice. In addition, stakeholders help to determine the expected levels of proficiency, or standards of achievement, for each learning outcome.

The content and structure of the Syllabus were motivated, in part, by an understanding of how it is to be used. The Syllabus, customized with results of stakeholder surveys, lays the foundation for curriculum planning and integration, teaching and learning practice, and outcomes-based assessment. Three goals motivated the choice of content and structure of the Syllabus. These goals were to:

- Create a structure whose rationale is clearly visible
- Derive a comprehensive high-level set of goals that correlate with other respected sources
- Develop a clear, complete, and consistent set of topics to facilitate implementation and assessment

The point of departure for the derivation of the content of the Syllabus is the simple statement that engineers *engineer*, i.e., they build products, processes, and systems for the betterment of humanity. In order to enter the contemporary profession of engineering, students must be able to perform the essential functions of an engineer. As described previously, graduating engineers should be able to conceive–design–implement–operate complex value-added engineering products, processes, and systems in a modern, team-based environment. Stated another way, graduating engineers should appreciate the engineering process; be able to contribute to the development of engineering products, processes, and systems; and do so while working in engineering organizations. Implicit is the additional expectation that, as university graduates and young adults, engineering graduates should be mature and thoughtful individuals.

These high-level expectations map directly to the first-level, or X-level, organization of the Syllabus, as illustrated below. The mapping of the first-level Syllabus items to the four expectations illustrates that a mature individual interested in technical endeavors possesses a set of Personal and Professional Skills and Attributes, central to the practice. In order to develop complex, value-added engineering systems, students must have mastered the fundamentals of the appropriate Technical Knowledge and Reasoning. In order to work in modern, team-based environments, students must have developed the Interpersonal Skills of teamwork and communication. Finally, in order to create and operate products, processes, and systems, students must understand something of CDIO Systems in the Enterprise and Societal Context.

As shown below, the CDIO Syllabus is organized at the first two levels in rational manner. The first level, or X level, reflects the functions of an engineer, who is a well-developed individual, involved in a process that is embedded in an organization, with the intent of building products, processes, and systems. The second level of detailed content, or X.X level, reflects contemporary practice and scholarship of the engineering profession and, in the list below, are related to ABET EC 2000 Criteria 3a–3k.

The CDIO Syllabus

1. Technical knowledge and reasoning
    - 1.1 Knowledge of underlying sciences [a]
    - 1.2 Core engineering fundamental knowledge [a]
    - 1.3 Advanced engineering fundamental knowledge [k]
2. Personal and professional skills and attributes
    - 2.1 Engineering reasoning and problem solving [e]
    - 2.2 Experimentation and knowledge discovery [b]
    - 2.3 System thinking
    - 2.4 Personal skills and attributes
    - 2.5 Professional skills and attitudes
3. Interpersonal skills: Teamwork and communication
    - 3.1 Teamwork [d]
    - 3.2 Communications [g]

3.3 Communication in foreign languages

4. Conceiving, designing, implementing, and operating systems in the enterprise and societal context

   4.1 External and societal enterprise [h]
   4.2 Enterprise and business and context
   4.3 Conceiving and engineering systems [c]
   4.4 Designing [c]
   4.5 Implementing [c]
   4.6 Operating [c]

The Syllabus is further defined to third- and fourth-levels of detail, respectively, the X.X.X level and the X.X.X.X level. These fine-grain details are necessary to transition from high-level goals to teachable and assessable learning outcomes. Although it could seem overwhelming at first, the detailed Syllabus has many benefits for engineering faculty who may not be experts in some of the Syllabus topics. The details provide insight into content and learning outcomes, the integration of these skills into a curriculum, and the planning of teaching and assessment. The above list is a condensed version of the CDIO Syllabus at the second level of detail. The complete CDIO Syllabus can be found in Rethinking Engineering Education: The CDIO Approach (Crawley et al. 2007; or http://www.cdio.org).

## The CDIO Standards

As noted above, evaluation is defined as a process for determining the merit or worth of a program by comparing the evidence collected to some set of expectations. In the CDIO standards-based program evaluation approach, programs are compared to an explicit set of expectations, namely the 12 CDIO Standards. In the CDIO approach, evaluation information is used to help make programs better by guiding the allocation of resources for improvement. A CDIO standard describes an essential characteristic of an engineering program that has adopted the CDIO approach to engineering education. The 12 standards were developed in response to requests from industrial partners, instructional program leaders, and alumni for the specific attributes of CDIO programs, i.e., they wanted to know how they would recognize CDIO programs and their graduates. As a result, the CDIO Standards:

- Define the distinguishing features of a CDIO program
- Serve as guidelines for educational program reform
- Create Quality Assurance benchmarks and goals that can be applied world-wide
- Provide a framework for self-evaluation and continuous improvement

Taken individually, the CDIO Standards add little new knowledge to the literature on effective engineering education practice. However, taken as a whole, the 12 CDIO Standards provide a comprehensive approach to the reform and improvement of engineering programs.

The 12 CDIO Standards that follow address program philosophy (Standard 1), curriculum development (Standards 2, 3, and 4), design–implement experiences and workspaces (Standards 5 and 6), new methods of teaching and learning (Standards 7 and 8), faculty development (Standards 9 and 10), and assessment and evaluation (Standards 11 and 12).

## *Standard 1: The Context*

Adoption of the principle that product, process, and system lifecycle development and deployment – Conceiving, Designing, Implementing and Operating – are the context for engineering education.

## *Standard 2: Learning Outcomes*

Specific, detailed learning outcomes for personal, interpersonal, and product, process, and system building skills, as well as disciplinary knowledge, consistent with program goals and validated by program stakeholders.

## *Standard 3: Integrated Curriculum*

A curriculum designed with mutually supporting disciplinary subjects, with an explicit plan to integrate personal, interpersonal, and product, process, and system building skills.

## *Standard 4: Introduction to Engineering*

An introductory course that provides the framework for engineering practice in product, process, and system building, and introduces essential personal and interpersonal skills.

## *Standard 5: Design–Implement Experiences*

A curriculum that includes two or more design–implement experiences, including one at a basic level and one at an advanced level.

### *Standard 6: Engineering Workspaces*

Workspaces and laboratories that support and encourage hands-on learning of product, process, and system building, disciplinary knowledge, and social learning.

### *Standard 7: Integrated Learning Experiences*

Integrated learning experiences that lead to the acquisition of disciplinary knowledge, as well as personal, interpersonal, and product, process, and system building skills.

### *Standard 8: Active Learning*

Teaching and learning based on active experiential learning methods.

### *Standard 9: Enhancement of Faculty Skills Component*

Actions that enhance faculty competence in personal, interpersonal, and product, process, and system building skills.

### *Standard 10: Enhancement of Faculty Teaching Competence*

Actions that enhance faculty competence in providing integrated learning experiences, in using active experiential learning methods, and in assessing student learning.

### *Standard 11: Learning Assessment*

Assessment of student learning in personal, interpersonal, and product, process, and system building skills, as well as in disciplinary knowledge.

### *Standard 12: Program Evaluation*

A system that evaluates programs against these 12 standards, and provides feedback to students, faculty, and other stakeholders for the purposes of continuous improvement.

Full descriptions and the rationale for each of the CDIO Standards may be found in Rethinking Engineering Education: The CDIO Approach (Crawley et al. 2007; or http://www.cdio.org). For each standard, the description explains the meaning of the standard; and the rationale highlights reasons for setting the standard.

## Program Evaluation and Quality Assurance Aligned with the CDIO Standards

As illustrated in Fig. 1, evaluation of a CDIO program focuses on the objectives and outcomes of the program and the processes that contribute to students' achieving them, within the context of the institutional mission and program goals. These processes include: the curriculum and its related syllabus (based on the CDIO Syllabus), teaching and learning methods, the learning environment, learning assessment, and faculty development. Note that program evaluation is itself one of the standards and therefore one of the processes that contributes to the accomplishment of the program objectives and outcomes. The key Quality Assurance questions aligned with the CDIO Standards are listed here. These questions might be applied to any program in any discipline.

- What are the objectives and outcomes of a CDIO program? How are they aligned with institutional mission and program goals? What is the context for these objectives and outcomes? (Standards 1 and 2)
- How does a CDIO curriculum contribute to the attainment of program outcomes? How are CDIO outcomes embodied in the CDIO Syllabus integrated into the curriculum? (Standard 3)

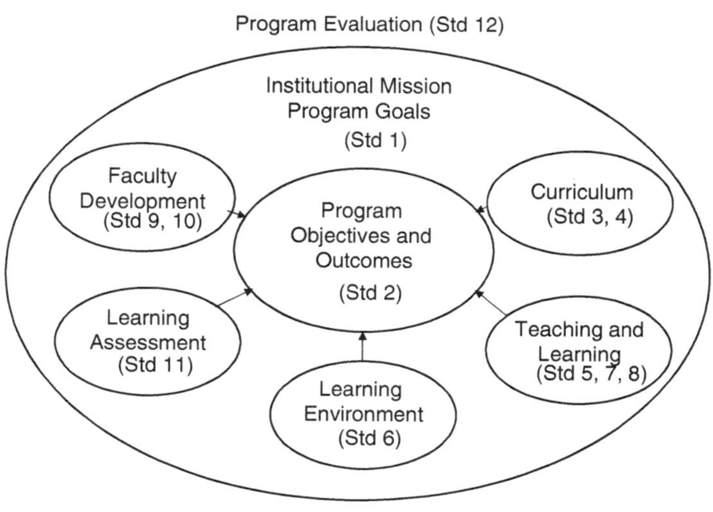


**Fig. 1** Program evaluation and Quality Assurance aligned with the CDIO standards

- How do first-year courses introduce the CDIO context and motivate students to choose engineering programs? (Standard 4)
- How do active and experiential methods contribute to the attainment of program outcomes in a CDIO context? How are these learning experiences integrated into the engineering program? (Standards 5, 7, and 8)
- How does the learning environment contribute to the attainment of CDIO program objectives and outcomes? (Standard 6)
- What have students achieved with respect to program outcomes? How are CDIO learning outcomes measured and documented? (Standard 11)
- How are faculty development and motivation encouraged? How do faculty roles change in a CDIO context? How satisfied are faculty with the teaching and learning experiences? (Standards 9 and 10)
- Is there a systematic process in place to evaluate CDIO program outcomes and processes? Are the evaluation results used in continuous process improvement? (Standard 12)

## CDIO Standards and Engineering Education Quality Assurance

The CDIO program evaluation approach expands the Quality Assurance criteria of ABET EC2000 particularly in the areas of teaching and learning, and the consequent need for faculty development. Table 1 provides a comparison of the CDIO Standards and the ABET evaluation criteria set forth in EC2000. A CDIO program recognizes that a shift in focus, context, and outcomes requires support for instructional staff. If faculty members are expected to integrate all CDIO learning outcomes into their courses, they need to enhance their own experiences in them. And if new program outcomes require new methods of teaching, learning, and assessment, instructors need support to make these changes, as well. A CDIO program evaluation examines the nature and level of support that is provided to the entire instructional staff.

### *Continuous Program Improvement*

Self-evaluation provides opportunities to not only rate current status, but also plan specific actions for continuous program improvement. These steps for improvement are aligned with the CDIO Standards. For example, a Quality Assurance plan would implement steps to:

- Monitor the integration of CDIO learning outcomes into each course and revise course plans where necessary (Standard 3)
- Monitor and support capstone courses, and investigate ways to include more Implement and Operate experiences into the program (Standard 5)

**Table 1** CDIO Standards compared with ABET's *EC2000*

| CDIO Standard | EC2000 (ABET) |
|---|---|
| 1. Adoption of the principle that product, process, and system lifecycle development and deployment – Conceiving, Designing, Implementing, and Operating – are the context for engineering education | No explicit statement, but implicitly engineering science is the context |
| 2. Specific, detailed learning outcomes for personal and interpersonal, and product, process, and system building skills consistent with program goals and validated by program stakeholders | Criterion 2a.Detailed published educational objectives that are consistent with the mission of the institution and these criteria<br>Criterion 2b. A process based on the needs of the program's various constituencies in which objectives are determined and periodically evaluated |
| 3. A curriculum designed with mutually supporting disciplinary subjects, with an explicit plan to integrate personal and interpersonal, and product, process, and system building skills | Criterion 2c. A curriculum and process that ensures the achievement of the program objectives<br>Criterion 4. A general education that complements technical content of curriculum and is consistent with program and institution objectives |
| 4. An introductory course that provides the framework for engineering practice in product, process, and system building, and introduces essential personal and interpersonal skills | Not addressed |
| 5. A curriculum that includes two or more design–implement experiences, including one at a basic level and one at an advanced level | Criterion 4. Curriculum to culminate in a major design experience, based on knowledge and skills acquired in earlier coursework |
| 6. Workspaces and laboratories that support and encourage hands-on learning of product, process, and system building skills, disciplinary knowledge, and social learning | Criterion 6. Classrooms, labs, and equipment must be adequate to accomplish program objectives, foster faculty–student interaction, and encourage student professional development |
| 7. Integrated learning experiences that lead to the acquisition of disciplinary knowledge, as well as personal and interpersonal, and product, process, and system building skills | Not addressed |
| 8. Teaching and learning methods based on active experiential learning models | Not addressed |
| 9. Actions that enhance faculty competence in personal and interpersonal, and product, process, and system building skills | Criterion 5. General requirements for faculty competence, but no explicit requirement for system–building skills |
| 10. Actions that enhance faculty competence in providing integrated learning experiences, in using active experiential teaching and learning methods, and in assessing student learning | Not addressed |

(continued)

**Table 1** (continued)

| CDIO Standard | EC2000 (ABET) |
|---|---|
| 11. Assessment of student learning in personal and interpersonal, and product, process, and system building skills, as well as in disciplinary knowledge | Criteria 3a–3k. An assessment process to demonstrate that graduates have developed a set of specific attributes and abilities, listed in a–k |
| 12. A system that evaluates programs against the 12 standards, and provides feedback to students, faculty, and other stakeholders for the purposes of continuous improvement | Criterion 2d. A system of ongoing evaluation that demonstrates achievement of program objectives, and uses the results to improve the effectiveness of the program |

- Investigate new sources of challenging design problems (Standards 3, 5, and 7)
- Incorporate the experiences and best practices of successful instructors (Standard 8)
- Make connections from professional development activities to more effective student learning and satisfaction (Standard 10)
- Expand the set of tools for assessing CDIO learning outcomes, and extend the use of these tools to a greater number of courses (Standard 11)
- Close the loop on data collection and process improvement (Standard 12)

## Summary

Two of the important contributions of the CDIO approach to engineering education include the CDIO Syllabus and the CDIO Standards. The CDIO Syllabus defines the learning requirements, i.e., the knowledge, skills, and attitudes that a graduating engineer should possess. The definition, validation, and integration of these learning outcomes are included in the program evaluation framework as Standard 2.

The 12 standards developed by the CDIO Initiative serve as a useful framework for internal program self-evaluation and external Quality Assurance. Chalmers University of Technology, the Royal Institute of Technology, Linköping University, and the Massachusetts Institute of Technology have been using this model of self-evaluation since October 2000. New collaborators – more than 35 engineering programs world-wide – conduct similar self-evaluations as they begin their reform process, and as they project their desired status in 2–5 years.

The CDIO Standards are consistent with Quality Assurance criteria in the United States, Canada, the United Kingdom, and South Africa. In Sweden, for example, academic groups responsible for the evaluation of higher education programs have adopted the CDIO Standards as the basis of their evaluation processes. With its emphasis on continuous program improvement, the CDIO standards-based approach adds value to program review, accreditation and other Quality Assurance schemes.[2]

[2] For a more extensive discussion of CDIO and national Quality Assurance in Sweden, see the chapter "Quality Assurance of Engineering Education in Sweden" by Malmqvist and Sadurskis.

## References

Accreditation Board for Engineering and Technology. (2000). *Engineering Criteria 2000.* http://www.abet.org

Brodeur, D. R., Crawley, E. F., Ingemarsson, I., Malmqvist, J., & Östlund, S. (2002). International collaboration in the reform of engineering education. *Proceedings of the 2002 American Society of Engineering Education Annual Conference and Exposition, Montreal, Canada.* The American Society of Engineering Education (ASEE) holds the copyright to papers published in the proceedings. http://www.asee.org/conferences/v2search.cfm

Crawley, E. F., Malmqvist, J., Östlund, S., & Brodeur, D. R. (2007). *Rethinking engineering education: The CDIO approach.* New York, NY: Springer.

# Using Soft Systems Thinking to Confront the Politics of Innovation in Engineering Education

**Henk Eijkman, Obada Kayali, and Stephen Yeomans**

**Abstract** Engineering curriculum innovators face a range of formidable barriers which, singly or in combination, have thwarted countless attempts at sustainable curricular quality improvement initiatives regardless, of their educational efficacy. The often ignored *elephant in the room* of programmatic quality improvement is the politics of change. The essential point of this chapter is this: a whole-of-programme curriculum innovation demands an intervention strategy capable of effectively responding to multiple stakeholder perspectives and therefore to the politics of change. It is argued that Soft Systems Methodology embedded within a Systemic Action Research approach will give engineering educators that capability.

## Introduction

Engineering and engineering education are embedded in rapidly evolving economic, social and professional environments (National Academy of Engineering 2005). Therefore systemic curriculum innovation is no longer an optional extra but a key component of continuous quality improvement in engineering education. Moreover in a highly competitive global higher education environment curriculum, innovation represents a high stake and, for some, a high risk activity. The barriers to systemic whole-of-programme innovation are all too evident. As the National Academy of Engineering (2005, p. 13) points out,

> Scattered interventions across engineering education over the past decade or so have not resulted in systemic change, but rather only in isolated instances of success in individual programmes, on individual campuses.

H. Eijkman (✉)
The University of New South Wales at The Australian Defence Force Academy, Bldg 28 Rm 204, Canberra ACT 2600, Australia
e-mail: h.eijkman@adfa.edu.au

A.S. Patil and P.J. Gray (eds.), *Engineering Education Quality Assurance: A Global Perspective*,
DOI 10.1007/978-1-4419-0555-0_18, © Springer Science+Business Media, LLC 2009

All of this makes curriculum innovation worthy of attention as a valuable sub-system of a larger continuous engineering education Quality Assurance system.

The essential thesis of this chapter is this: large-scale curriculum innovation demands a convergence of divergent stakeholder perspectives. Therefore to achieve sustainable, whole-of-programme, curriculum innovation we need an intervention strategy that has the capacity to effectively respond to the politics of change. Our argument is that soft systems thinking and in particular Soft Systems Methodology (SSM) provides engineering educators with that capacity.

We take this stance because we consider curriculum innovation, especially at programme level, to be a high-stake, high-risk component of the total Quality Assurance system in engineering education. While whole-of-programme innovation presents academic institutions and faculty (teaching academics) with substantive opportunities, they tend to be subject to considerable challenges and often deep-seated resistance. Documented accounts of substantive programme renewals invariably focus on the (invariably successful) introduction of new educational ideas and practices. Yet these accounts of change seldom if ever discuss how they have managed to resolve a host of often deeply entrenched organisational, cultural, philosophic and practical tensions and considerations.

Engineering curriculum innovators face a range of formidable barriers which, singly or in combination, have thwarted countless attempts at sustainable curricular quality improvement initiatives regardless, or even in spite, of their educational efficacy. The generally ignored *elephant in the room* of programmatic quality improvement is the politics of change. Seldom does the literature delve into the realities of convincing a range of diverse stakeholders how to come to an agreement on adopting new educational ideas. This is particularly pertinent given that many have divergent and strongly held opinions and will often do all in their power to protect their perceived interests. Accounts of innovation failure are, understandably, seldom if ever published although much can be learned from them. Published case studies of curricular innovation invariably consign the politics of change to the margins if mentioned at all.

At this point it is important to clarify that by *politics* we refer to power arrangements that shape relations between stakeholders with different interests and the ability (or not) to influence decision making and resource allocation. In this chapter therefore we examine large-scale curriculum innovation as a process that is primarily about the ability (or not) to influence decision making and allocate resources. Following in the spirit of the National Academy of Engineering we too are not interested in scattered and isolated initiatives. Hence in this chapter the term *innovation* refers specifically to systemic change at the level of a programme or degree course.

A review of the literature and our own experience indicates an endemic absence of any serious consideration of the politics of change. This is evident in curriculum innovation in the field of higher education generally and in the *hard* sciences, such as engineering, more particularly. In engineering education, power, social relations and cultural change management are mostly seen as *soft* issues that belong to the realm of the *social* rather than the *hard* sciences.

In this chapter we cannot, due to the limitations of space, present engineering educators with a detailed step-by-step *curriculum change manual*. What we will do is provide a framework capable of addressing the oft-neglected politics of curriculum innovation. In so doing, we provide a roadmap to guide academic leaders to achieve sustainable innovation in engineering curricula. Accordingly this chapter has two aims. Our first aim is to locate curriculum quality improvement within its broader social and political context as the rationale for a systemic, soft systems driven, action-focused change management strategy. Our second aim is to outline the key features of this soft systems strategy for guiding programme-level curriculum innovation. While different approaches are available (see Midgley 2000), we specifically focus on the application of SSM. We conclude by proposing that our SSM driven approach can be further enhanced when situated within a Participatory Action Research (PAR) continuous improvement spiral.

## Curriculum Innovation as a Socio-Political Process

Our first task is to convince engineering educators that curriculum change, while having a *hard* technical (educational content) component, is always situated within a *soft* social context. This social context is characterised by stakeholders who hold a variety of perspectives and interpretations and whose interests may differ considerably. Therefore achieving curricular innovation is mostly about managing a contentious and conflict-ridden socio-political process. This is why we must frame programmatic innovation as purposeful intervention within a complex set of *soft*, that is, *human* activities.

Making the distinction between hard and soft systems thinking and situating curriculum innovation within a soft systems approach is pivotal. Engineering education reform efforts – as for instance manifest in the CDIO methodology (see chapter "CDIO and Quality Assurance: Using the Standards for Continuous Programme Improvement" by Brodeur and Crawley), is dominated by hard systems thinking. This *conceptual reframing* of curriculum innovation provides the rationale for our SSM-driven framework. SSM has the inherent capacity to address the socio-political context of change. Its capacity to respond to the socio-politics of curriculum intervention is based on a critical distinction between hard (designed) and soft (human) systems thinking and practice, an issue to which we now turn.

Curriculum innovation is subject to converging as well as divergent top-down, bottom-up, and push–pull forces. Innovation can, for example, be pushed from the top by institutional demands or by recommendations from audits by professional bodies, or be instigated from below by enthusiastic faculty or a mix of both convergent and divergent push–pull forces. Either way programmatic curriculum change, especially when focused on a substantive transformation, comprises intervention in one or more social systems, such as funding arrangements, institutional/professional roles and expectations. As such, innovation does not merely constitute a *technical* shift in an educational direction. However, the engineering education literature rarely, if ever alerts, change agents to the need to view the curriculum design process as an inherently political procedure that demands a high level of competence in organisational change management. When it does refer to the politics of change it is done as a warning but without actually providing practical guidance.

In the final analysis, the literature portrays curriculum change as a rational process primarily informed and shaped by debates about educational *technicalities* and thus abstracted from the political realities of organisational change management. Yet whole-of-programme innovation immerses many people with different world-views, values and expectations, let alone levels of educational expertise, in a complex purposive activity. The starting point for innovation therefore is not to see this solely as technical–rational task involving "instrumental problem solving made rigorous by the application of scientific theory and technique" (Schon, 1983, p. 21) but rather to see it as a purposeful and contentious human activity in which everything, from values and perceptions to educational options, may all be *up for grabs*.

Hence, as many change agents have themselves discovered, programmatic curriculum transformation is a sensitive, difficult, messy and problematic process. Even the soundest proposals for educational innovation have to negotiate power arrangements and conflicting and often deeply entrenched organisational, professional, and personal priorities, values and interests. We argue, therefore, that even when reports of successful large-scale curriculum change projects fail to mention the political dimension (see for example Crosthwaite et al. 2006) that success is ultimately due to a team's ability to effectively drive the politics of change. The key to innovation is dynamic leadership; the willingness and ability to get multiple stakeholders to collectively champion an educational transformation *that everyone can live with.*

This is borne out in a study by Gruba et al. (2004) in a survey of 19 institutions in Australasia. They found that curriculum change is influenced not so much by academic motives or educational merit but is driven or restrained by macro- (institutional) and micro-level (personal) political factors. These include institutional considerations, outspoken individuals, and practical concerns such as student demand. Similarly a study by Sunal et al. (2001) identified a range of barriers to change. These included perceived lack of resources, especially time; turf conflicts; personal resistance to change and lack of training. The point is that higher education curriculum innovation literature still focuses on the internal or structural organisation of a curriculum. Even if there is an awareness of power and politics, the *political issues* focus on the form, content and function of the curriculum itself and not on the politics of the actual change process (see for instance Pinar et al. 1996; Bocock 1994; Jones 2002 and, for an exception, Arnold 2004).

In general the engineering education curriculum literature fares a little better. There are at least some engineering educators who recognise that curriculum innovation lacks models for guiding the actual curriculum change process (Froyd et al. 2000, 2006; Merton et al. 2001). For example, Froyd et al. (2000) at least put forward a curricular change model predicated on the need to build a cohesive coalition around any innovation proposal. Early evaluation results indicate that this organisational change model is effective in achieving systemic and sustainable curriculum innovation. Yet its dissemination and implementation by the engineering education community seems to have been quite limited (Froyd et al. 2000, see also Millar et al. for the results of using a similar process).

Naivety about organisational and cultural change management remains the Achilles heel of successful programme-level curriculum innovation. As pointed out by Froyd et al. (2006), this lack of awareness about, and inability to harness, organisational change management results in a slow rate of curricular reform and a high failure rate in achieving sustainable deep-seated curricular innovations. As McWilliam et al. (2008) point out, any successful reorganisation of an area of human activity, such as curriculum innovation, is very much predicated on the capacity of the re-organisers to engage relevant stakeholders in such projects.

The question is how do we build the capacity of engineering educators to engage with the politics of change? Extending the work of Kotter (1996) and Froyd et al. (2000, 2006), we consider the role of systems thinking. More explicitly we propose the inclusion of SSM as an effective framework to guide sustainable curriculum innovation.

## From Hard Systems Engineering to Soft Systems Thinking

How well are engineers and engineering educators placed to adopt a systems approach to programmatic intervention? Engineers, and indeed engineering educators, are generally well versed in hard systems engineering. This involves the ability to construct a well-defined system to achieve specified objectives. One example is designing a system to supply a town with potable water within a defined timeframe and budget. This mechanistic way of thinking depends largely on a foundational or realist epistemology; an understanding of our knowledge as an objective representation of the real world (Eijkman 2008).

However, as engineering researchers at the University of Lancaster discovered, a hard systems engineering approach is well suited to deal with *hard* clearly structured, engineering-oriented problem situations. Yet such a hard systems approach proved utterly unsuited to dealing with very messy and complex human situations which lack both clear problem definitions *and* agreed objectives. As observed by Checkland and Scholes (1990, p. 17) the hard systems engineering approach,

> is predicated on the fact that the need and hence the relevant need-meeting system, can be taken as given. Systems engineering looks at *how to do it* when *what to do* is already defined.

Many attempts to apply hard engineering systems thinking to soft human/social problems have ended problematically. This has led practitioners to conclude the need for a shift from *hard* to *soft* systems thinking (Checkland 1981). Again, by *hard* we refer to a systems-based methodology, such as systems engineering (Hall 1974; Daenzer 1976) in which the objective to be achieved is clearly specified and is therefore taken as given. A *soft* system on the other hand is also a systems-based methodology but one in which the objective to be achieved is generally obscure and, therefore, cannot be easily defined, let alone easily agreed upon and met (Checkland 1981). Table 1 identifies some key differences between these two positions.

The unsuitability of *hard* systems engineering to messy ill-defined human problem situations has led to a distinct shift from *hard* (mechanistic) to *soft* (human) systems thinking characterised by an epistemology in which our knowledge of the world is socially constructed (Checkland 1981; Checkland and Scholes 1990; Eijkman 2008). Therefore, it is soft systems thinking that "generates and works with an evolving appreciation of people's points of view and intentions" (Flood 2001, p. 137). It follows that sustainable programmatic curriculum innovation needs above all to be seen as intervention in a human system. This requires that change agents understand about emergent systems of meaning and dilemmas that surface in the process of wanting to introduce innovations within a real-world engineering education programme.

If curriculum innovation is to be successful and sustainable it needs to be approached holistically. One way is to adopt a soft systems perspective as a means to address, in one form or another, the interconnected hopes and concerns of all stakeholders. These may inevitably involve issues such as departmental funding arrangements, cultural issues around roles and rewards, priorities, union concerns around workloads, individual perceptions about the efficacy of change, students' perceptions about assessment loads, as well as any technical arguments about the engineering content and the structure of the educational innovation itself.

**Table 1** The hard and soft systems traditions (From Checkwell and Howell 1998)

| | *Hard* systems thinking | *Soft* systems thinking |
|---|---|---|
| Concept of organisation | Social entities that identify and seek to achieve goals | Social entities which seek to manage relationships |
| Underpinning systems thinking | The world is assumed to be systemic.<br>Solving problems means choosing between alternative means of achieving a known goal<br>A system is then engineered to achieve the stated objective | A process of inquiry that itself constitutes a system.<br>Solving problems means tackling ill-structured, hard to define problems in which the goal cannot be taken as given<br>A system-based means of structuring a debate |
| Underpinning methodology | Positivist, hypothesis testing, quantitative, modelling | Interpretative, action-research oriented, aims for gaining insight and understanding, qualitative, participative |

So far we have explained in broad brush strokes how soft systems thinking is an effective way to address the complexities of curricular quality improvement initiatives. Let us now, within the limited confines of this chapter, briefly describe what the use of soft systems thinking might actually involve.

## A Soft Systems Approach to Curricular Innovation

The critical issue in the complex interactions among stakeholders that make curriculum innovation such a complicated and contentious process is one of 'boundaries'. The first step is to discover and respond to the stakeholders involved and reflect on their viewpoints and interests (Midgley 2000). Stakeholders, when given the opportunity, inevitably present a wide range of issues which demonstrate that the boundary of intervention extends far beyond the *educational technicalities* of change. Moreover such issues are not *value free*.

It is clear that to be successful an intervention strategy needs to deal with these institutional, cultural, personal, as well as educational boundary issues holistically and systemically. This reflects the core concepts that underpin systems thinking, namely *emergence* and *interconnectedness* (Checkland 1981; Midgley 2000; Flood 2001). Applied to an engineering education programme *as a system*, *emergence* and *interconnectedness* mean that we can only make sense of resistance to curriculum innovation when we understand resistance as the emergent property of interconnected stakeholder positions, *as a whole*, rather than as the property of the constituent elements (the particular groups or individual stakeholders) of this curriculum system.

Hence we can only begin to fully appreciate the reasons for resistance and/or acceptance of curriculum innovation and, thereby, respond accordingly to facilitate change, only when we see an engineering education programme as an interconnected *human activity system*. That is, an engineering curriculum is a human activity system with its own culture made up of individuals with differing institutional, professional, technical, educational, and personal perspectives, interests and preferences. Additionally we also need to recognise that local curricular *systems* and their cultures are embedded in other, wider, systems. This is where larger issues related to *boundary judgments* come into play. This culture-centred perspective allows us to see the emergent properties of curriculum innovation as a whole. This is not possible if we focus on its component parts – for example educational technologies, engineering content, pedagogical methods, individual stakeholders, etc. – in isolation.

In summary, soft systems thinking is of considerable value as a problem structuring methodology for achieving sustainable curriculum innovation. A soft systems approach to managing the innovation process enables leaders in engineering education to:

> conceptualize complex situations characterized by interacting issues and multiple, conflicting viewpoints; reflect on values and boundaries of inclusion, exclusion and marginalization (of stakeholders and issues); sweep into intervention the viewpoints of a wide variety of stakeholders, including those who find themselves marginalized; and choose and/or design

> methods that provide the means to engage with others in a flexible and responsive manner, thereby facilitating the development of new social agendas and plans for change that can command widespread support from those affected by them. (Midgley 2000, p. 16).

## Soft Systems Methodology as the Approach of Choice

Having made the case for the appropriateness of *soft* rather than *hard* systems thinking in curriculum change management, we now describe SSM itself. SSM is based on a well-established set of theories and methodologies (see for example, Midgley 2000, 2001; Jackson 1990; Flood and Jackson 1991; Rosenhead 1989; Reason and Bradbury 2001). SSM engages participants iteratively, usually within a series of facilitator-led workshops, in a seven-step process of structured activities involving group discussions. The process of using SSM in a curriculum innovation, simplified for the purposes of brevity, can be summarised as follows:

1. Participants representing all stakeholders (see step 3 below) explore the curriculum change proposition in an open unstructured form. This encourages them to express their ideas and issues and captures the multiple perspectives and understandings about the innovation.
2. Participants express their multiple understandings by way of a rich visual map that provides a detailed picture of the current situation. Such a map also makes visible the interconnections between the various aspects of the proposed change.
3. Participants now explore key questions about the proposed innovation. This enables the convergence of various viewpoints and harmonises understandings. The aim is to identify potentially *relevant systems* that could be designed to implement an agreed-upon innovation process. The curriculum transformation process is conceptualised by defining relevant systems using the CATWOE mnemonic. In a curriculum innovation process, CATWOE would be operationalised as follows:
   C (customers) = students
   A (actors) = faculty (teaching academics)
   T (transformation process) = the conversion of current curriculum into new curriculum
   W (worldview) = The worldview which makes the curriculum change meaningful in its specific context
   O (owners) = The university and faculty – whose representatives can promote or hinder innovation
   E (environmental constraints) = Elements outside the system that impact on the proposed innovation but which are a "given" e.g. funding arrangements, the demands of professional bodies etc.
4. Participants now identify, link and prioritise conceptual models or maps of the activities to be undertaken. This enables them to operationalize each system (identified in step 3) needed to implement the innovation.

5. Here participants compare their conceptual models with the real world of the proposed curriculum innovation (as expressed in the rich picture drawn up in step 2) in order to evaluate the feasibility of their models.
6. Next participants, given the crucial role of cultural and political imperatives that govern the change process, critically analyse the proposed innovation process in terms of what is systemically desirable and culturally feasible. As Checkland and Scholes (1990, pp. 52–53) point out, those "intellectually locked within the hard paradigm, believing the world to be systemic, will imagine that changes have to be systemically feasible and culturally desirable." At the conclusion of this step participants produce an action plan for implementing, monitoring and evaluating the innovation so as to have an evidence base for reflection and ongoing improvements.
7. Participants are now in a well-informed position to proceed with the implementation of the innovation. One approach, as described below, is to follow a PAR process. Regardless of the implementation process followed, stakeholders know that what they have all worked towards is not just systemically desirable but most importantly culturally feasible (Checkland and Scholes 1990; Checkland and Howell 1998).

At this point it is important to point out that when participants become familiar with this approach they are increasingly comfortable in adapting SSM. As Checkland and Howell (1998, p. 162) note, SSM "as a methodology – a set of principles of method rather than a precise method – has to be adapted by its users both to the demands of the situation they face and to their own mental modes and casts of mind".

To conclude our description of this systemic process of structuring curriculum innovation, we also propose that consideration be given to situating SSM within a PAR framework.

## Locating SSM within a PAR framework

The effectiveness of our proposed framework for guiding programmatic curricular innovation can be enhanced by positioning SSM within a broader PAR approach. There is a considerable history of the application of action research to education, especially in English speaking countries (Kemmis and Grundy 1997; Zeichner 2001). In addition, PAR is particularly suited to researching the implementation of educational innovations in higher education. This is particularly so in institutions which purport to support *student-centred learning*. This is because the specific focus on the *participatory* aspect of action research explicitly promotes the inclusion of all stakeholders. This ensures the inclusion of students. This is of particular interest to those of us who are very much aware that the voice of students is notably absent in many if not most curriculum change initiatives (Eijkman et al. 2005).

For our purposes, and building on Wadsworth (1998) we can define PAR as "an approach to curriculum action research that involves all relevant stakeholders in all facets of the change process – from planning to evaluation and reflection on its actual implementation – in order to ensure that all who have a stake in the curriculum are actively involved in the process of reshaping it". The inclusion of a PAR approach acts as a counter to curriculum innovation initiatives when these are imposed top-down or lack meaningful consultation and/or student involvement.

The inclusion of an overarching PAR framework also ensures that the change process is inherently reflective and focused on outcomes that are open to continuous improvement. While an extended discussion is not possible here, suffice it to say that this framework guides participants through a robust continuous improvement spiral. This spiral consists of *planning, acting, monitoring/evaluation, reflection and renewed planning for the next cycle.* Hence the PAR process is always responsive to issues encountered in the implementation process (Kemmis and McTaggart 1988). In this regard, a further case can also be made for constituting faculty into communities of curricular practice, but we shall leave that for another time and place as well.

## Conclusion: Towards Systemic Action Research

In this chapter, we have put the case for a curricular intervention strategy capable of successfully engaging the politics of culture change. This is critical if we are to optimise our chances of achieving large-scale and sustainable programme-level curriculum innovation. In making this case we have proposed that SSM, especially when located within a wider PAR framework, provides an innovation capacity building strategy par excellence.

We propose that this framework, which for convenience we term *Systemic Action Research* (SAR), extends the work of Kotter (1996) and Froyd et al. (2000, 2006). It provides a sound systemic process to guide engineering educators through the steps of establishing a need, gathering a cohesive leadership team, agreeing on learning outcomes, etc. These are complex and messy social activities, which themselves benefit from a more rigorous understanding and structuring. But simply identifying these steps, though necessary, is insufficient. This is why we believe that SAR's convergence of SSM and PAR has much to offer the curriculum quality enhancement process.

SAR has the capacity to effectively drive the ongoing planning, action, evaluation, reflection and refinement spiral in curricular innovation. The SAR guide to curriculum intervention places the cultural and political context of curriculum innovation squarely at the centre of the innovation process. It provides a robust and proven methodology to guide a collaborative approach to continuous quality improvement in programmatic curriculum innovation.

The SAR framework takes account of, and is responsive to, the complexities of institutional, professional and personal boundaries and power arrangements in which curriculum innovation operates. The end result is that we not only enhance the curriculum under consideration but, because the SAR process also captures knowledge about the improvement process, we engage in a *capacity-building* process as well. We look forward to the ways in which readers may adapt this approach to innovation in engineering education in their own institutional settings.

SAR, therefore, enhances the quality of both the *product*, i.e., curriculum innovation, and *process*, i.e., planned change, by including stakeholder collaboration in all aspects of the ongoing curricular innovation spiral. The SAR team involves all stakeholders in planning and implementation, monitoring, evaluation, reflection and ongoing refinement. In this way SAR, as a comprehensive outcomes-driven guide to curriculum innovation, represents a key Quality Assurance mechanism in engineering education.

## References


Ackoff, R. L. (1981). *Creating the corporate future*. New York, NY: Wiley.

Arnold, G. B. (2004). Symbolic politics and institutional boundaries in curriculum reform: The case of National Sectarian University. *The Journal of Higher Education, 75*(5), 572–593.

Bocock, J. (1994). Curriculum change and professional identity: The role of the university lecturer. In J. Bocock & D. Watson (Eds.), *Managing the university curriculum: Making common cause* (pp. 116–126). Bristol, PA: Society for Research into Higher Education & Open University Press.

Brodeur, D., Crawley, E., Ingemarsson, I., Malmqvist, J., & Ostlund, S. (2002). International collaboration in the reform of engineering education. *Proceedings of the 2002 American Society for Engineering Education Annual Conference & Exposition.*

Checkland, P. B. (1981). *Systems thinking, systems practice*. Chichester: Wiley.

Checkland, P. B., & Howell, S. (1998). *Information, systems and information systems*. Chichester: Wiley.

Checkland, P. B., & Scholes, J. (1990). *Soft systemsmethodology in action*. Chichester: Wiley.

Crawley, E. F. (2002). *Creating the CDIO syllabus, a Universal Template for Engineering Education*. Paper presented at the 32nd ASEE/IEEE Frontiers in Education Conference, November 6–9, 2002, Boston, MA.

Crosthwaite, C., Cameron, I., Lant, P., & Litster, J. (2006). Balancing curriculum processes and content in a project centered curriculum in pursuit of graduate attributes. Reprinted from *Education for Chemical Engineers, 1*, 39–48, Trans IChemE, Part D, 2006.

Daenzer, W. F. (ed). (1976). *Systems engineering*. Cologne: Peter Hanstein.

Eijkman, H. (2008). Web 2.0 as a non-foundational network-centric learning space. *Campus-Wide Information Systems, 25*(2), 93–104.

Eijkman, H., Backhouse, J., & Holgate, C. (2005). *Quality as a homely place for the subaltern: The promise of post-colonial action research in transcultural universities*. Workshop for the International Practitioner Research Conference & Collaborative Action Research Network Conference, 4–6 November, Utrecht, The Netherlands.

Flood, R. L. (2001). The relationship of 'systems thinking' to action research. In P. Reason & H. Bradbury (Eds.), *Handbook of action research: Participative inquiry and practice* (pp. 133–144). London: Sage.

Flood, R. L., & Jackson, M. C. (1991). *Creative problem solving: Total systems intervention*. Chichester: Wiley.

Froyd, J., Penberthy, D., & Watson, K. (2000). Good educational experiments are not always good change processes. *30th ASEE/EEE Frontiers in Education Conference, October 18–21, Kansas City, MO.*

Gruba, P., Moffat, A., Sondergaard, H., & Zobel, J. (2004). What drives curriculum change? In *Conferences in Research and Practice in Information*. Paper presented at the Sixth Australasian Computing Education Conference, Dunedin, New Zealand.

Hall, A. D. (1974). Three dimensional morphology of systems engineering. In F. Rapp (Ed.), *Contributions to a philosophy of technology*. Dordrecht: Reidel.

Jackson, M. C. (1990). Beyond a system of systems methodologies. *Journal of the Operational Research Society, 39*, 715–724.

Jones, E. A. (2002). Transforming the curriculum: Preparing students for a changing world. *ASHE-ERIC Higher Education Report, 29*(3), 124.

Kemmis, S., & Grundy, S. (1997). Educational action research in Australia: Organisations and practices. In S. Hollingsworth (Ed.), *International action research: A casebook for educational reform* (pp. 40–48). London: Falmer press.

Kemmis, S., & McTaggart, R. (1988). *The action research planner* (3rd ed.). Geelong: Deakin University.

Kotter, J. (1996). *Leading change*. Boston, MA: Harvard Business School Press.

Mason, R. O., & Mitroff, I. I. (1981). *Challenging strategic planning assumptions*. New York, NY: Wiley.

McWilliam, E., Hearn, G., & Haseman, B. (2008). Transdisciplinarity for creative futures: What barriers and opportunities. *Innovations in Education and Teaching International, 45*(3), 247–253.

Merton, P., Clark, C., Richardson, J., & Froyd, J. (2001). Engineering curricular change across the foundation coalition: Potential lessons from qualitative research. *Proceedings, Thirty-First ASEE/IEEE Frontiers in Education Conference, 3, Reno, NV, F4B15–20.*

Midgley, G. (2000). *Systemic intervention: Philosophy, methodology, and practice*. New York, NY: Kluwer.

Millar, S. B., & Courter, S. S. (1996). From promise to reality: How to guide an educational reform from pilot stage to full-scale implementation. *Prism, 6*, 31–34.

Morecroft, J. D., & Sternman, J. D. (Eds.). (1994). *Modeling for learning organizations*. Portland, OR: Productivity Press.

National Academy of Engineering. (2005). *Educating the engineer of 2020. Adapting engineering education to the new century*. Washington, DC: The National Academies.

Newble, D., & Cannon, R. (2001). *A handbook for medical teachers* (pp. 89–108). London: Kluwer.

Pinar, W. F., Reynolds, W. M., Slattery, P., & Taubman, P. M. (1996). *Understanding curriculum: An introduction to the study of historical and contemporary curriculum discourses*. New York, NY: Peter Lang.

Reason, P., & Bradbury, H. (2001). *Handbook of action research: Participative inquiry and practice*. London: Sage.

Rosenhead, J. (1989). *Rational analysis for a problematic world*. Chichester: Wiley.

Schon, D. A. (1983). *The reflective practitioner: How professionals think in action*. London; Temple Smith.

Sunal, D. W., Sunal, C. S., Whitaker, K. W., Freeman, L. M., Odell, M., Hodeges, J., et al. (2001). Teaching science in higher education: Faculty professional development and barriers to change. *School Science and Mathematics, 101*(5), 246–257.

Wadsworth, Y. (1998). What is participatory action research? *Action Research International*. Paper 2. Available online at http://www.scu.edu.au/schools/gcm/ar/ari/p-ywadsworth98.html

Zeichner, K. (2001). Educational Action Research. In P. Reason & H. Bradbury (Eds.), *Handbook of action research: Participative inquiry and practice* (pp. 273–283). Sage: London.

# Real-Time Quality Control Methods in PBL-Based Engineering Education

**Egon Moesby and Palle Qvist**

**Abstract** Traditionally post-semester evaluations are most often used to measure and report higher education quality. The demands for Quality Assurance, quality control and quality development due to the Bologna Declaration and subsequent institutional and programme accreditation and audit demands will expand the use of evaluations. However, traditional post-semester standardised, summative evaluations are not an effective way to monitor quality as the students' progress in their studies through the semester. Instead, there is a need for a system that enables the institution to measure quality and performance of the instruction and learning by the students as they occur, and to respond accordingly. The School of Basic Studies at Aalborg University, Denmark, operates such a system. The chapter describes the setup of a real-time evaluation system that allows students to have a democratic influence on their education.

## Introduction

Traditionally, *post-semester* evaluations are most often used to measure and report higher education quality. It is one thing, however, to fulfil external quality demands by conducting standardised post-semester evaluations, which are generally used to *produce quantity performance indicators*. It is quite another to have a system that enables the institution to measure the quality and performance of the instruction and learning by the students as they progress, and to respond accordingly.

Based on experience from the School of Basic Studies at Aalborg University, Denmark, such post-semester performance methods and indicators are not an

E. Moesby (✉)
School of Basic Studies, The Faculties of Engineering, Science and Medicine,
Aalborg University, Strandvejen 12-14, DK-9000 Aalborg, Denmark
e-mail: hem@aaue.dk

A.S. Patil and P.J. Gray (eds.), *Engineering Education Quality Assurance: A Global Perspective*,
DOI 10.1007/978-1-4419-0555-0_19, 

effective way to monitor the quality as the students' progress in their studies *through* the semester. Instead, there is a need for a system that assesses the *real-time* quality of teaching and learning and provides feedback for improvement. It is through such a system that *students* can have a genuine and direct influence on their educational situation through the semester, which makes it possible for students to actively participate in monitoring and managing their own learning.

The Schools of Basic Studies of Engineering, Science and Medicine house the first year of studies for all the programmes offered by the Faculties of Engineering, Science and Medicine. The school run the first year programmes at campus Aalborg, campus Copenhagen and campus Esbjerg.

The 2007 intake was approximately 1,100 students distributed among the following studies:

- Architecture and Design
- Civil- and structural Engineering
- Energy
- Electronic Systems
- Data Engineering
- Product- and Design Psychology
- Medialogy
- Natural Science
  - Mathematics
  - Physics
  - Chemistry
  - Data Science
- Nanotechnology
- Biochemistry
- Surveying
- Mechanical Engineering
- Bachelor IT
- Global Business Development
- Geography
- Environmental Planning
- Health Technology
- Medicine, Industrial Specialisation
- Sports Science

The overall *organisational structure* for first year studies is shown in Fig. 1 and the *structure of the first year programme* is shown in Fig. 2.

The arrangements for Quality Assurance (QA) described here are based on several years of experience, development and refinement. Although the Quality Assurance system is not claimed to be complete, it is coherent and had been shown to serve its purpose very well. Consequently it might inspire others in their efforts of designing or in improving a quality programme.

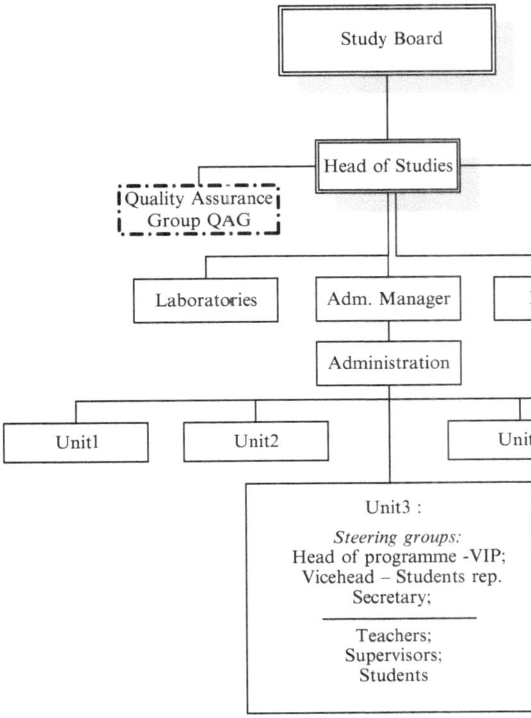


**Fig. 1** Organisational structure for first year studies

## The Real-Time Quality Assurance Setup

To fulfil the aim of assessing students' progress and the quality of instruction and, if necessary, taking real-time action to improve the quality of instruction, the Study Board and the Head of Study depend on synchronised feedback from students, teachers, laboratory staff and the administration. In order to acquire such data, there need to be, first, a structure for giving feedback and, second, confidence on the part of all involved that they can experience benefits from the results.

With reference to the organisational structure in Fig. 1, the Study Board is the legal body and the Head of the Study Board is the Head of Study as well. The Head is assisted by the administration and a Study Assistant. Further support for the studies is provided by the Information Technology unit laboratories and workshops that also report to the Head. The students and staff are organised into units, which entail a number of programmes. These units are locally administered by a group of secretaries and a number of programme heads. In order to be able to administer a real-time evaluation programme, several additional units and related activities have been created.

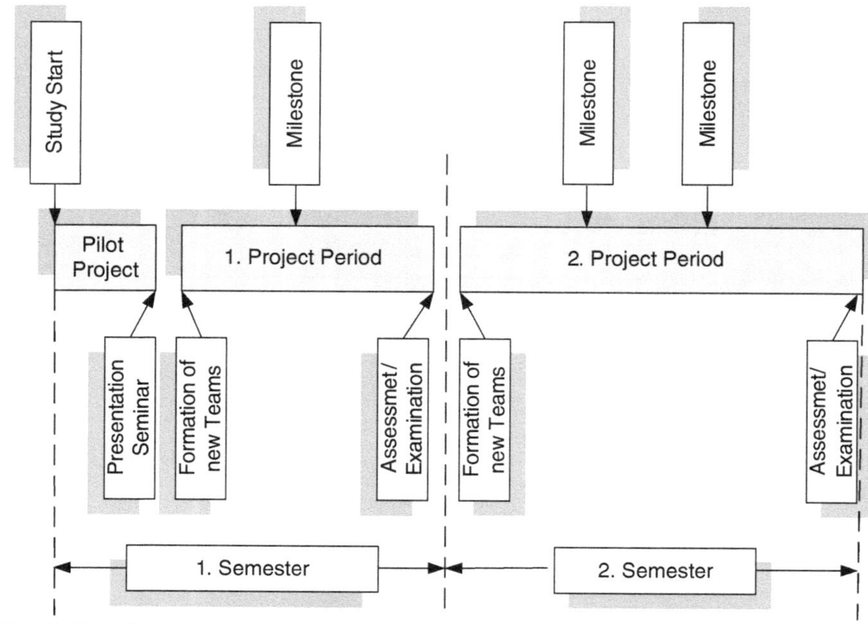


**Fig. 2** Overall structure of the first year programme

The total and complex quality control and quality development system constituted hereby is according to Stufflebeam, a *Management Information System* (Stufflebeam 2001). The units are administrative groups that include the students enrolled under the specific programmes within the unit, the supervisors from the respective programmes, teachers delivering subjects and a secretary who is responsible for the administration of the local unit. At present, the School of Basic Studies have three such units in Aalborg, one unit in campus Copenhagen and one unit in campus Esbjerg.

## Steering Group

Within each unit, a Steering Group is formed. Each steering group consists of one student representative from each team, the head of programme and teacher representatives. Ideally the student represents the opinions from the entire group of teams. The teacher representative is someone acting as a co-supervisor, dealing with study topics other than the professional content.

As a service to the groups and to ensure that all steering groups cover the same items of business, meetings follow a guide with a pre-defined agenda. Of course, topics other than those listed in the guide can be added, if needed. (One of the students

acts as secretary, and makes official minutes of the meeting, which are made public.[1]) The following is an example of a standard agenda:
Steering Committee Meeting No.: (Day, room, address, time)

1. Call to order
2. Roll call
3. Old business and approval of last meeting's minutes
4. Election of a student Vice-Head (only for the first meeting of 1 and 2 semester).
5. Announcements
6. Status of project work
7. Subject evaluations (a) Mathematics, (b) Cooperation, Learning and Project Management, (c) Technology, Mankind and Society, (d) Specific subject 1, (e) Specific subject 2, (f) Seeking of information, (g) Introducing of laboratories, (h) Free study activities and (i) Adjournments
8. Status of support (a) Information Technology department, (b) Service officers, (c) Administration and (d) Laboratories and (e) rooms in general
9. Next meeting(s)
10. Adjournment

Each programme is led by a programme head who is responsible to the Head of Study. The programme head is responsible for the daily conduct and administration of the study programme. The duties of the day-to-day administration are carried out with support from the unit secretary. The unit secretary further organises the planning, schedules and evaluation plans, in cooperation with the programme head. The programme head participates in the steering group's monthly meeting. The steering group's meetings are normally planned to be held prior to board meetings.

The vice-head for each steering group is elected by the group's student representatives. The Head of Study calls for two meetings during each semester to discuss matters directly with the steering groups' vice-heads. The topics addressed during these meetings include the quality control and quality development of ongoing teaching and supervision. Minutes are kept of the meetings and made public. In essence, this group serves as a focus group with respect to Quality Assurance.

The heads of programme participate in a meeting prior to study board meetings to discuss the groups' problems and progress with the Head of Study, the Board Secretary and the Study Assistant. The issues specifically addressed during these meetings include the quality control and quality development of the ongoing teaching and supervision; however, the discussions are not limited to these topics, as other issues are on the agenda as well.

---

[1] Restrictions: By demand from the Rector's office and from union representatives, the Study Board must ensure, prior to publication that the minutes are in accordance with rules, which imply that the minutes may not contain any offensive or discriminating content, and may not make reference to gender, race, colour, religion, economy, disability, etc. Further, no names must be published if related to criticisms. However, the original text will be made known for the Study Board. The Head of Studies or Study Assistant goes through all minutes from the steering groups, and if any corrections have to be made, it is sent back to the writer for correction.

## The Quality Assurance Group

The internal Quality Assurance Group (QAG) is the part of the organisation shown in Fig. 1 that continuously assesses the quality development in the study programmes and gives recommendations based on their findings to the Head of Study and Study Board. The QAG consists of three members appointed by the Head of Study. Normally two of the members are external in order to insure a high degree of objectivity. External members means the involvement of persons who are external to the programme or activity being judged.

The QAG group works according to terms of reference decided by the Study Board. The terms of reference describe the general topics that the QAG will address. In some situations, the Head of Study or Study Board may wish to have a specific area addressed, in such a situation the Head of Study defines the task and releases additional resources specifically for this task. The standard tasks the QAG undertakes are as follows:

- Production and publication of minutes from their meetings
- Post-semester evaluations
- Post-semester reports
- Quality checks among selected project reports and reflection documents and analysis of these; and comparing their findings with the Intended Learning Outcomes (ILO)
- Evaluating to what degree the given subjects can be found utilised in the project reports
- Based on the Steering Groups minutes, evaluating the judgements of the given subjects
- Each semester and after each educational year, publishing *a state of the education* report.

Examples of specific tasks undertaken by QAG include a survey of drop-out students (to uncover why they drop out and what they will do instead); study of the use of lap-top computers; a review of programmes with specific problems and variations in marks after a change in the evaluation methods.

Figures 3 and 4 show different ways of presenting output data from a post-semester evaluation. As a criterion of success, the Study Board set 70–75% of the combined very satisfied and satisfied answers as an acceptable level.

## Reflection-in-Action Loops

The milestone activities, shown in Fig. 2, are very important elements in the students' project work and learning process. It is an internal process for the group, but the lessons learned through the process can be found in steering group meeting minutes. The main reason for these activities is to require students to make reflections on where they have been, what they have done so far, where they are now and where they wish to go in the rest of the project period.

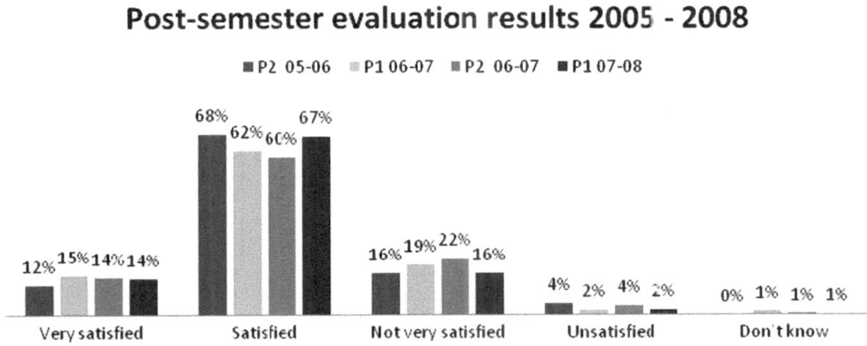


**Fig. 3** Overall evaluation results for the period 2005–2008 (QAG School of Basic Studies of Engineering 2008, p. 7)

| **Judgement of question: Introduce students to scientific work and scientific practices with special focus on methods, theories and models** | Procentage of answers | Procentage of answers, sum |
|---|---|---|
| Very satisfied | 23 | |
| Satisfied | 54 | 77 |
| Less satisfied | 21 | |
| Unsatisfied | 2 | **23** |
| Don´t know | 0 | 0 |
| Total | 100 | 100 |

**Fig. 4** Overall evaluation with summarised clusters of pleased and not pleased student answers (QAG School of Basic Studies of Engineering 2007, p. 11)

These reflection activities, or so-called Reflection-in-Action Loops (Cowan 2006), are just the reflections on their doings within the project in progress. One milestone activity is scheduled in the first project period, and two in the second project period. The content of these milestone activities is described in terms of the elements that need to be addressed.

The reflection-in-action loops are based on the setup shown in Fig. 5. The basis for the examination is shown at the left side of the dotted line. Beside the regulations stated in the Study Guide, the Project Report should include documentation of the extent to which the ILO have been fulfilled. In addition, in order to make explicit the actual learning outcomes attained in relation to a specific project from the student's point of view, a post-semester Reflection Document is required. In this document, students compare their observations of real-time experiences with a reflection at the end of the semester about their overall learning.

This reflection document contains self-evaluations of the taught subjects, the cooperation with teachers and supervisors, the relevance of the subjects and

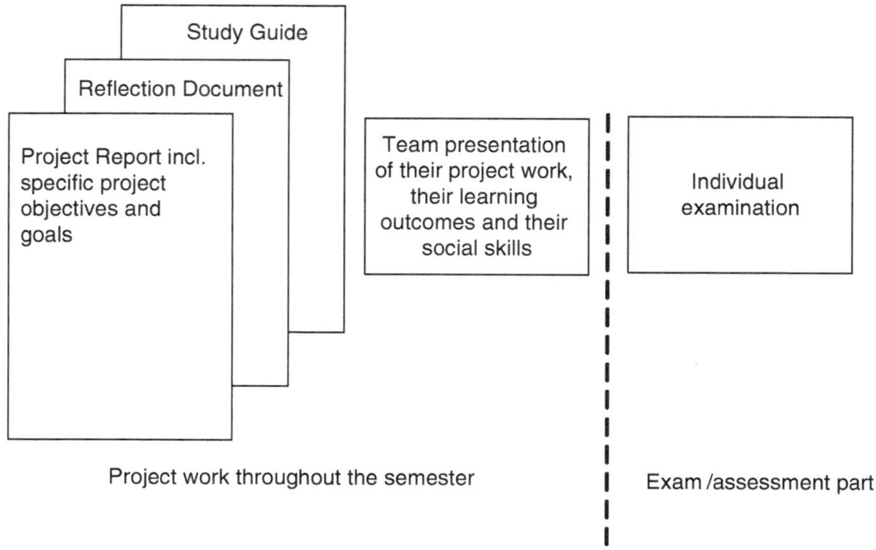


**Fig. 5** The setup of a project examination

the attainment of the ILO. It further contains a section relating to the students' learning outcomes as individuals, as well as describing their social competences. The document relies on the team's diary or journals; within the process, one and a half days are allocated to the writing of this document. A teacher gives an introductory lecture for the activity, and makes a short summary of the disciplines taught relating to the area. The students write the reflection document after handing in the project report. The reflection documents provide invaluable data that form a central role when the QAGs conduct their evaluations and make their reports.

## Management

A decentralised Study Counselling Office (SCO), run by the Study Board with students employed as officers, contributes valuable information for the real-time evaluation. They keep statistics, and report if any anomaly in students' behaviour occur when visiting the SCOs office. They take part in the Study Board meetings as observers and give updated information to the Board. Further they have close contact with the Head of Study and administrators who provide valuable input for the assessment of the quality as well.

The issues of the daily administrative performance and Quality Assurance are permanent topics on the agenda of the regular meetings of the administrative staff. There are two kinds of meetings. One meeting deals with the standard administration issues and any problems that are occurring. The second kind concerns overall performance and involves all categories of staff, including administrators, IT-support staff and laboratory staff. Minutes are kept of the meetings, and are published. Having regular meetings helps to insure that any irregularities, anomalies, problems, etc. are identified and dealt within a very short time. During the meetings, any cross-sectional problems are also addressed; and if any section has problems with students, for example, other sections will learn of them and be able to take any necessary action.

Informal interaction in small circles and casual meetings is as important as formal meetings. The authors believe strongly in the importance of having close contact with many people, through small discussions and casual meetings, as the information gained is often very valuable. Also such information can be spread faster in the organisation in this way, compared with the formal meeting process. As a result, networks becomes tighter, which in turn enhances the possibility for swift action when problems arise, which the students appreciate very much.

While it may be true that the documentation of quality in real-time evaluations lacks precise evidence as it is quite formative in nature, the post-semester evaluations make up for this by making more summative and exact judgements of the quality of work during the semester based on the evaluation of the final outcomes regarding student performance.

Using the perspective of the learning organisation (Argyris 1993), the daily quality performance-assurance activities can be said to be a single-loop-learning setting and the post-semester evaluations are a double-loop-learning situation. The post-semester evaluations are a double-loop-learning situation since they take results of the day-to-day single-loop-learning activities and add another loop for synthesis and reflection. Single-loop learning guides day-to-day administrative performance, Quality Assurance and problem solving, whereas double-loop learning guides changes of a higher magnitude dealing with more fundamental issues such as improvements in the organisation or in the educational practices and traditions. Figure 6 illustrates the Quality Assurance activities just described aligned with a timescale.

In summary, the real-time activities, while being part of an overall Quality Assurance policy, are formative and constitute a progressive process in which the students, teachers, supervisors and administration can actively participate. The post-semester evaluations are retrospective and adjudicating (Qvist and Moesby 2008), i.e. summative as far as the current cohort is concerned, since there are no possibilities for students to be actively involved in subsequent activity.

From the students' perspective, the real-time evaluation activities are a forum for having actual influence on their ongoing situation which the following example shows:

> Some students may not be pleased with the performance of a teacher when giving lectures. What do they do? According to the standards, which have been made public and discussed in one of the first steering group meetings, the students will address the teacher directly to try to solve the problem locally. In most cases this contact should solve the problem, and no further action will be needed. However, in some cases, this is not sufficiently helpful, and so the situation will be raised during a steering group meeting, and be discussed in the presence of the Programme Head. Further, the episode is described in the minutes from the meeting, and thus made known to the administration and Head of Study. The procedure

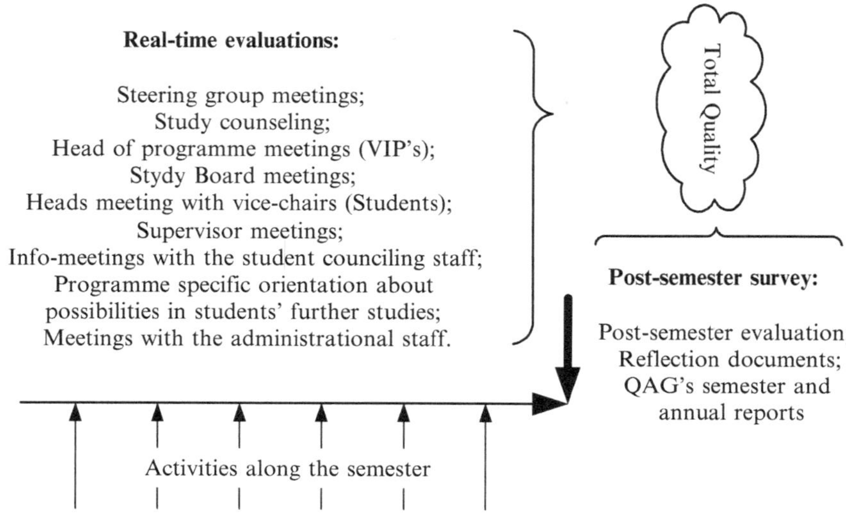


**Fig. 6** QA activities for a semester

then is that the Programme Head will make contact with the teacher to try to solve the situation. In some cases this will solve the problem, but – in rare cases – it will not. In such cases, the students now have the possibility to make contact in a formal letter to the Head of Study, who then will deal with the problem. Typically, by having a meeting with the teacher, and in case that this is not giving the expected result, then a formal complaint is sent to the Head of Department where the teacher is based. In one situation, this process has been carried out in a period of approximately two weeks, and resulted in substituting the teacher. This was a radical solution, but it serves to illustrate, how the students can see that their problem is taken seriously, that the "system" is supportive and that they feel that they have real influence on their situation.

A post-semester evaluation may indeed contain valuable and useful assessments and suggestions for improvements. However, it will not be able to improve the situation for the students involved, as they have just finished their semester. The result can – hopefully – be beneficial for the next cohort of students entering the semester.

## Discussion

Since 2006, the post-semester evaluations and other surveys have been conducted electronically. Prior to this the evaluation forms were sent out to the students by mail with the data being entered and results calculated manually. By using electronic forms, much time has been saved, and the results can be published as soon as the survey period is over. In some situations, it is even possible to see the results in real time, where the output data is updated daily.

One of the biggest problems we have with the post-semester evaluations is the low frequency of responses. The percentage of responses is approximately 30% of the total population, and there are big differences in the frequencies between the three campuses. The low frequency is likely to be the result of the voluntary participation in the surveys. Some institutions have the answering of the post-semester evaluation integrated as a part of the students' work, and in some situations they even give credit which is added to a summarised grade.

The low frequency means that the answers are not representative. On the other hand, we tend to believe that those students answering the questions are those students who have opinions about their situation. We further tend to think that the answers might be a bit on the negative side, as students who wish to express dissatisfaction with their situation are more likely to answer than the students who might be quite content with their situation.

Moreover, having the results and the real-time evaluations as an indicator of the situation during the semester, we believe it somewhat compensate for the low response frequency of final evaluations. Further, the evidence from the real-time evaluations forms part of the QAG's work, together with the post-semester evaluations; so in this respect, the total results of the two main evaluation methods are assessed and evaluated in the QAG reports to the Study Board.

## Conclusions

As described above the numbers of separate evaluations employed by that School of Basic Studies at Aalborg University, Denmark is quite high. To a great extent this is the case because of their utility in providing real-time formative data as well as input for summative evaluations. Another reason is the expectations embodied in the Bologna declaration that include the public availability of evaluation data.

The Bologna declaration says that in order to promote the European system of higher education world-wide, European countries should promote European co-operation in Quality Assurance with a view to developing comparable criteria and methodologies. Chapter "Quality Assurance in European Engineering Education: Present and Future Challenges" (p. 11) by Cowan makes a significant observation: "The keyword, perhaps, is comparable – which does not necessarily mean *identical*". That is a point that should not be forgotten when working to meet demands and requirements from the outside. Demands for comparability are not limited to Europe as is illustrated by the other chapters in this book, as the same processes are being developed all over the world in cultures where the approach to the making and publication of judgements can vary markedly.

It is the authors' hope that this chapter has shown how a real-time evaluation environment can be constructed and administered. If such a system was not present, it would have been difficult to react swiftly to any anomaly or dissatisfaction within the programme or among the students. As it is, many smaller conflicts and potential problems may be averted. These are problems that might otherwise develop into quite serious crises later on.

Such an environment also encourages students to be active participants in a process that aims to improve the quality of teaching and their learning as they go along. This allows them to have a strong and real democratic influence on their education. What the students find most important is the short response time within the system to problems raised. However, the environment for the teachers and supervisors is also improved, as problems are solved relatively early in the process, and can then be a part of a learning process for all instead of simply being a topic for complaint in the post-semester evaluations.

## References

Argyris, C. (1993). *Knowledge for action*. San Francisco, CA: Jossey-Bass.

Cowan, J. (2006). *On becoming an innovative university teacher. Reflection in action* (2nd ed.). Maidenhead: Open University Press, McGraw-Hill Education.

QAG School of Basic Studies of Engineering, S. a. (2007). *QAG P1 report 2006–07*. Aalborg: GSN, Aalborg University.

QAG School of Basic Studies of Engineering, S. a. (2008). *QAG P2 report 2007–08*. Aalborg: GSN, Aalborg University.

Qvist, P., &. Moesby, E. (2008). *Quality assurance in a PBL learning system in engineering and science education*. Paper presented at the SEFI 36th Annual Conference July 2–5 2008, Alborg University, Denmark. Accessed 21 December 2008. http://tnb.aau.dk/~pal/kug/1037_qvist_moesby.pdf

Stufflebeam, D. L. (Ed.). (2001). Evaluation models. In *New directions for evaluation* (Vol. 89). San Francisco, CA: Jossey-Bass.

# Enhancing the Quality of the Engineering Student Experience

**Chenicheri Sid Nair and Arun Patil**

**Abstract** Institutions of higher education all over the world are abuzz with concerns over quality, Quality Assurance, continuous quality improvement and rankings of quality. These concerns have been stimulated by increasing internationalisation and globalisation, increasing student enrolments, the expansion of distance and e-learning education, changing funding patterns for higher education, etc. Students' feedback is an integral part of the continuous quality improvement and assurance process. This chapter reports the process of designing and implementing the Monash Experience Questionnaire which was used to gather student views of their experience while studying engineering at Monash. The lessons learned from this process helped the authors to identify best practices regarding the role of student opinion surveys in Quality Assurance schemes.

## Introduction

Despite the fact that Quality Assurance of educational environments is both a complex and a subtle enterprise, remarkable progress has been made in conceptualising and assessing determinants of quality. For example, studies have consistently confirmed a strong correlation between the quality of classroom environments and student learning and satisfaction (Devlin 2002; Fraser 1991, 1994, 1998; Suarez et al. 1998; Nair and Fisher 2001; Ramsden 1991). In addition, research over the last four decades has shown that students' and teachers' perceptions of quality are important social and psychological aspects of the educational environments (Fraser 1991, 1994, 1998). This research points to the importance of collecting reliable and

A. Patil (✉)
Faculty of Sciences, Engineering and Health, CQUniversity Australia, Mackay Campus, 12 G.03, Boundary Road, Mackay QLD 4741, Australia
e-mail: a.patil@cqu.edu.au

A.S. Patil and P.J. Gray (eds.), *Engineering Education Quality Assurance: A Global Perspective*,
DOI 10.1007/978-1-4419-0555-0_20, 

valid data on students' and teachers' perceptions in order to monitor and, if necessary, improve the quality of the teaching and learning environment.

Apart from data that are routinely available such as enrolment patterns, retention and completion rates and grade distributions, there is a need to document and understand student perceptions about their experiences. Research shows that evaluations of students' perceptions are not only reliable and valid indicators of course and programme quality but also useful sources of information about ever-changing cohorts of students. In addition, there is clear evidence that feedback from students' evaluations can lead to improved teaching effectiveness thus enhancing the quality of the educational environment (Marsh 1987, 2001; Marsh and Dunkin 1997; Marsh and Roche 1993).

In short, the lessons learned from the meta-analyses of research undertaken over the last 40 years provide strong evidence that student evaluations are a valid and reliable means of assessing teaching effectiveness and the quality of the educational environment in general. In particular, student evaluation surveys provide (Bennett et al. 2006):

- Diagnostic feedback about teaching that are useful in the development and improvement of teaching strategies
- Research data to underpin further design and improvement of units, courses and curricula
- A measure of teaching effectiveness that may be used in decision making such as performance management and appraisal
- Information helpful to current and potential students in selecting units, programmes and courses
- A measure for judging the quality of the education environment related to units, programmes and courses and institutions.

Although there has been substantial research undertaken on the effectiveness of student feedback, questions are sometimes raised about their validity (Feldman 1990, 1997; Marsh 2001; Nasser and Fresko 2006). One common criticism is that students lack the wisdom and experience to provide effective feedback. However, research shows a high correlation between student course-end ratings and ratings of instruction by peers, administrators and alumni (D'Apollonia and Abrami 1997; Marsh and Dunkin 1997; Marsh and Roche 1997; Overall and Marsh 1980).

In addition, the assertion that student feedback is just a popularity contest has been shown not to be the case as has the assertion that to gain good evaluations, teachers should simply make the course easy (Feldman 1997; Greenwald and Gillmore 1997; Marsh 1984, 1987). In fact, research demonstrates that teachers who assign more and difficult work tend to be rated as more effective teachers (Greenwald and Gillmore 1997; Marsh 1984, 1987).

In short, when the total volume of research on students' evaluation of the educational environment is considered, especially the meta-analyses of this research, it is clear that student feedback provides a valid and reliable assessment of teaching and learning outcomes. And, as such, it provides a relevant basis for making decisions about quality and its improvement.

## QA in Engineering Education and Students Feedback

The integration of student feedback into the Quality Assurance process of academic programmes is very common for professional courses such as engineering. Student feedback can be used not only to assess the current quality of courses, but also to guide the improvement of engineering classroom and laboratory practices and facilities as well as the overall quality of the engineering education environment. In addition, student feedback, if heeded can enhance institutional prestige in the competitive global educational marketplace (Nair and Patil 2008).

The literature on higher education Quality Assurance contains various advanced student feedback instrument and methods that promise better, faster and more cost-effective means of assessing and improving teaching and learning. In engineering education, the Conceive–Design–Implement–Operate (CDIO) initiative is a prominent example of a systematic quality improvement effort where Quality Assurance is monitored through rigorous student assessment and programme evaluation (see chapter "CDIO and Quality Assurance: Using the Standards for Continuous Programme Improvement" by Brodeur and Crawley) where student feedback plays a prominent role. In other example, Monash University have recently introduced a new Survey Management System (SMS) which enables quicker collection, analysis and reporting of student and staff survey data. It also centrally integrates a wide range of evaluation data available in the institution and enables more effective reporting according to particular needs, for instance, individual faculties or campuses across the University (Monash University 2009).

As with all good Quality Assurance systems there is a need to validate student feedback using other sources. This could include collecting data from a variety of student feedback tools which measure similar areas as well as from other stakeholders such as former students, employers and post-graduate institutions. For example, a research study carried out at the University of Cape Town in South Africa that investigated engineering graduates' perception about their readiness at work revealed that engineering graduates felt well prepared for industry with adequate expertise in technical skills. However, they identified their weaknesses in other important skills such as working in multi-disciplinary teams, leadership, practical preparation and management skill (Martin et al. 2005). Information from graduates is an important element in Quality Assurance. This triangulation process lends greater credibility to the information gathered. Therefore, it is essential to monitor and reflect upon the full spectrum of *student* feedback in order to devise and implement the best Quality Assurance mechanism in Engineering Education.

## The Monash Experience Questionnaire

The Monash Experience Questionnaire (MEQ) was developed in response to a recommendation of the 2002 institutional review report Still Learning: The Report of our Self Review (Monash University 2002). That report recommended "an ongoing

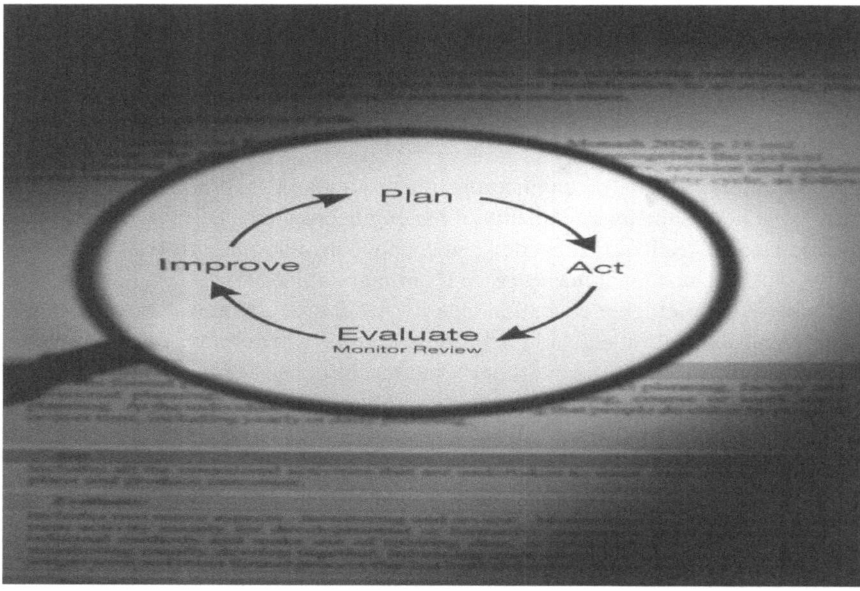


**Fig. 1** The quality cycle implemented at Monash University

Monash Student Experience Questionnaire as a means for determining, monitoring, benchmarking and improving the student experience" (p. 16). The MEQ is an important component of the Monash Quality Cycle – plan, act, evaluate (monitor and review) and improve, as outlined in Fig. 1.

Monash University's current approach to quality was articulated in 2001, in line with emerging national and international trends. It is based on the fitness-for-purpose understanding of quality. Monash sees quality as a cyclical endeavour that is everyone's responsibility, and uses internal feedback and external referencing to guide improvement (Monash University 2001).

The MEQ is noteworthy for two reasons. First, it provides a means for systematically identifying areas of success as well as areas that require attention from the University. This information is used in campus and faculty planning processes and in preparing for audits by external agencies such as the Australian University Quality Agency (AUQA). Second, the MEQ is a key indicator of student perceptions related to the Course Experience Questionnaire (CEQ), a national survey, which is mailed to each student completing an Australian undergraduate qualification both in Australia and offshore campuses. The results from the CEQ are used to make requests for additional funding for the University. In 2007, the University received over $10 million through the Learning and Teaching Performance Fund as a result of the information provided by the MEQ/CEQ.

## *Methodology Used*

The questionnaire designed by the core quality unit at Monash, the Centre of Higher Education Quality (CHEQ), includes seven scales containing 51 items plus two global satisfaction items. Each item in this questionnaire had the response alternatives of strongly agree, agree, neutral, disagree, strongly disagree and not applicable. Throughout, there is opportunity for students to comment on their experience. Descriptive information of each scale is shown in Table 1.

The other distinct areas measured in the questionnaire are the campus and the general university experience. These two dimensions comprise a total of 11 items. Descriptive information of these two scales is shown in Table 2.

All items were personalised, which means that students were asked for their personal perception of their role in the environment rather than their perception of the learning environment in the class as a whole (Nair and Fisher 2000).

**Table 1** Description of scales used to measure overall study and general university experience in the MEQ

| Scale | Items | Description |
|---|---|---|
| Good teaching (CEQ scale) | 6 | Measures student perception of teaching. It focuses on feedback, motivation, attention, understanding of problems and skill in explaining concepts |
| Generic skills (CEQ scale) | 6 | Measures student perception of generic skills (graduate attributes) development achieved in their courses |
| Learning community (CEQ scale) | 5 | Primarily focuses on student perceptions of the social experience of learning |
| Monash graduate attributes | 10 | Measures student perception of graduate attributes specifically identified at Monash |
| Student support/ resources | 9 | Measures student perception of study support and resources available at Monash |
| Monash approach to teaching and learning | 10 | Measures student perception of the particular approach to teaching and learning encouraged at Monash |
| Other important areas of teaching and learning | 6 | Measures student perception of key areas of teaching and learning not identified elsewhere (e.g., student workload) |

**Table 2** Description of scales used to measure the general campus and university experience

| Scale | Number of items | Description |
|---|---|---|
| General campus experience | 4 | Measures wider student experience at the campus in areas of student activities, social interaction and facilities |
| General university experience | 7 | Measures wider student experience at the University relating to opportunities to develop personally and academically, involvement in the decision making process, use of feedback and association with Monash |

## *Student Feedback and the Monash Insight*

The MEQ has provided insight into the engineering student experience by highlighting the stronger and weaker aspects of the learning environment. The results suggest that engineering students pursuing their studies at Monash are generally satisfied with their overall learning experience; however, at the same time bringing the attention of teachers within the university engineering community to some key areas that could make the learning experience much more positive. Monash as a whole has made significant progress in implementing a comprehensive student experience survey. Specifically it is one of the few universities to undertake such an extensive survey that has given insight to the engineering student experience as they progress through their studies.

A key element of any Quality Assurance process is the union of evaluation and improvement. It has been found that many organisations collect feedback but do not act upon the findings which suggest to customers that their views were not valued. It is best summarised in a paper presented by the Graduate Careers Council of Australia (GCCA) as follows (Institutional Arrangements for Student Feedback, Graduate Careers Council of Australia 1999, p. 20):

> It is a myth that all you have to do is to send back the result of a survey to those concerned and action, improvement and innovation will automatically occur. Such an assumption ignores all the research on motivation and change management in universities.

Powney and Hall (1998) suggest that in institutions where staff are not concerned about student opinion, student apathy toward the completion of feedback questionnaires is more apparent. Students are less likely to take the time and effort to complete questionnaire if they feel that it is a meaningless, resultless ritual that the institution goes through in order to meet Quality Assurance procedures. In fact, participants are reluctant to continue to provide feedback if there is little evidence of action taken in response to their feedback (Harvey 2003; Leckey and Neill 2001; Powney and Hall 1998). In this regard, Leckey and Neill (2001) argue that, closing the loop is the important issue in terms of total quality management. "If students do not see any actions resulting from their feedback, they may become sceptical and unwilling to participate" (p. 25). Harvey (2003) extends this argument with the qualification that not only action must take place as a result of student feedback, but also students need to be informed and convinced that change has occurred.

Monash like many universities has approached student evaluations very seriously. Student survey data provides strategic-level information while being student centred. This is exemplified by the quality process that has been developed. The MEQ has face validity within the Monash community and is used by the university as one of its performance indicators. For example, the utilisation of a systematic suite of student surveys at Monash has resulted in improved teaching, improved design of units, improved access to facilities like computer and other laboratories and improved communication between the faculty and the students.

Though there is considerable research that indicates that student survey response rates are declining and that participants are becoming more disengaged with surveys,

Monash has demonstrated that with proper engagement of students in the process results in response rates that are sufficiently high to be representative of the student cohort. This is supported by the response rates for unit evaluations that have risen from just over 30% in 2005 to close to 52% in 2008.

Over the three iterations of the MEQ there has also been a gradual increase in student satisfaction with the teaching and learning environment. To achieve this, the university, along with the respective faculties and staff, had to both understand the data and put forth actions plans designed to deliver the desired changes. This is the critical step of closing the loop in the quality cycle that is needed to make student evaluations effective tools for quality improvement.

## Conclusions

Clearly the survey results at Monash, along with the research literature in general, demonstrate that there is an advantage to engaging students, the key stakeholders in the educational environment, in the Quality Assurance and improvement process. Student surveys have in general provided institutions with a rich resource of students' perceptions of the teaching and learning environment.

By using student feedback to inform quality improvement efforts students will see that their opinions are valued by the institution. This is a critical factor in not only getting constructive feedback from the students but also sustaining their engagement. Monash has demonstrated the effectiveness of this form of engagement via the dramatic change in response rates.

In summary, an effective Quality Assurance scheme such as the one embodied in the Monash Quality cycle relies on the effectiveness of the student evaluation system. Most importantly there is a need for universities to realise that not only feedback from students is an important and integral part of the quality cycle but also such feedback provides reliable and valuable information on which a University can act to better meet the needs of its students.

## References

Bennett, L., Nair, C.S., & Wayland, C. (2006). *Love it or hate it: Participation a key ingredient in closing the loop*. Paper presented at the 2006 Australian Universities Quality Forum, Perth, Australia.

CDIO Initiative. (2005). http://www.cdio.org/

Cheng, Y. C., & Tam, M. M. (1997). Multi-models of quality in education. *Quality Assurance in Education, 5*, 22–31.

D'Apollonia, S., & Abrami, P. C. (1997). Navigating student ratings of instruction. *American Psychologist, 52*(1), 1198–1208.

Dee, K. C. (2004). *Reducing the workload in your class won't "Buy" you better teaching evaluation scores: Re-refutation of a persistent myth*. Paper presented at the ASEE Annual

Conference and Exposition, Salt Lake City, Utah, USA. http://md1.csa.com/partners/viewrecord.php?requester=gs&collection=TRD&recid=200502994CO&recid=2005078022969CE&recid=2005078029198MT&q=related%3AlJDKdf5nWGoJ%3Ascholar.google.com%2F&uid=1024956&setcookie=yes

Department of Education, Employment and Workplace Relations (DEEWR). (2008). *Views of engineering students: Report of a survey of final year university engineering students in Australia.* http://www.dest.gov.au/sectors/higher_education/publications_resources/profiles/views_engineering_students.htm

Devlin, M. (2002). An improved questionnaire for gathering student perceptions of teaching and learning. *Higher Education Research and Development, 21*(3), 289–301.

Feldman, K. (1990). Quoted in *The Teaching Professor, December 1990*, 5.

Feldman, K. (1997). Identifying exemplary teachers and teaching: Evidence from student ratings. In R. P. Perry & J. C. Smart (Eds.), *Effective teaching in higher education: Research and practice* (pp. 368–395). Bronx, NY: Agathon.

Fraser, B. J. (1991). Two decades of classroom environment research. In B. J. Fraser & H. J. Walberg (Eds.), *Educational environments: Evaluation, antecedents and consequences* (pp. 3–27). London: Pergamon.

Fraser, B. J. (1994). Classroom and school climate. In D. Gable (Ed.), *Handbook of research on science teaching and learning: A project of the National Science Teachers Association.* New York, NY: Macmillan.

Fraser, B. J. (1998). Science learning environments: Assessment, effects and determinants. In B. J. Fraser & K. G. Tobin (Eds.), *The international handbook of science education* (pp. 527–564). Dordrecht: Kluwer.

Greenwald, A. G., & Gillmore, G. M. (1997). No pain, no gain? The importance of measuring course workload in student ratings of instruction. *Journal of Educational Psychology, 89*, 743–751.

Harvey, L. (2003). Student feedback. *Quality in Higher Education, 9*(1), 3–20.

Institutional Arrangements for Student Feedback. (1999). *Graduate Careers Council of Australia.*

Leckey, J., & Neill, N. (2001). Quantifying quality: The importance of student feedback. *Quality in Higher Education, 7*(1), 19–32.

Marsh, H. W. (1984). Students' evaluations of university teaching: Dimensionality, reliability, validity, potential biases and utility. *Journal of Educational Psychology, 76*, 707–754.

Marsh, H. W. (1987). Students' evaluations of university teaching: Research findings, methodological issues, and directions for future research. *International Journal of Educational Research, 11*, 253–388.

Marsh, H. W. (2001). Distinguishing between good (useful) and bad workloads in student evaluations of teaching. *American Educational Research Journal, 38*(1), 183–212.

Marsh, H. W., & Dunkin, M. J. (1997). Students' evaluation of university teaching: A multidimensional perspective. In R. P. Perry & J. C. Smart (Eds.), *Effective teaching in higher education: Research and practice* (pp. 241–320). Bronx, NY: Agathon.

Marsh, H. W., & Roche, L. A. (1993). The use of students' evaluations and an individually structured intervention to enhance university teaching effectiveness. *American Educational Research Journal, 30*, 217–251.

Marsh, H. W., & Roche, L. A. (1997). Making students' evaluations of teaching effectiveness effective: The critical issues of validity, bias, and utility. *American Psychologist, 52*(11), 1187–1197.

Martin, S., Maytham, B., Case, J., & Fraser, D. (2005). Engineering graduate's perceptions of how well they were prepared for work in industry. *European Journal of Engineering Education, 30*(2), 167–181.

Monash University. (2001). *Quality at Monash: Values and principles.* http://www.adm.monash.edu.au/cheq/quality/

Monash University. (2002). *Still learning: The report of our self-review.* http://www.adm.monash.edu.au/cheq/reviews/still-learning/still-learning.pdf

Monash University. (2009). Centre for Higher Education Quality (CHEQ), http://www.adm.monash.edu.au/cheq/

Nair, C. S., & Fisher, D. (2000). Validation and application of a personalised form of a learning environment questionnaire in tertiary science classroom. In D. Fisher & J.-H. Yang (Eds.), *Improving classroom research through international cooperation* (pp. 419–428). Perth: National Key Centre for School Science and Mathematics.

Nair, C. S., & Fisher, D. L. (2001). Learning environments and student attitudes to science at the senior secondary and tertiary levels. *Issues in Educational Research, 11*(2), 12–31.

Nair, C.S., & Patil, A. (2008). Industry vs universities: Re-engineering graduate skills – a case study. *Proceedings of Australian Universities Quality Forum (AUQF2007), Canberra, Australia.*

Nasser, F., & Fresko, B. (2006). Predicting student ratings: The relationship between actual student ratings and instructors' predictions. *Assessment & Evaluation in Higher Education, 31*(1), 1–18.

Overall, J. U., & Marsh, H. W. (1980). Students' evaluations of instruction: A longitudinal study of their stability. *Journal of Educational Psychology, 72*, 321.

Powney, J., & Hall, S. (1998). *Closing the loop: The impact of student feedback on students' subsequent learning*. Edinburgh: Scottish Council for Research in Education.

Ramsden, P. (1991). A performance indicator of teaching quality in higher education: The Course Experience Questionnaire. *Studies in Higher Education, 16*(2), 129–149.

Suarez, M., Pias, R., Membiela, P., & Dapfa, D. (1998). Classroom environment in the implementation of an innovative curriculum project in science education. *Journal of Research in Science Teaching, 35*(6), 655–671.

# Taxonomies of Engineering Competencies and Quality Assurance in Engineering Education

L.C. Woollacott

**Abstract** This chapter reviews both literature and theory related to the identification and articulation of graduate attributes and competencies that are relevant to engineering education. Such attributes and competencies form the basis for Quality Assurance in engineering education. This chapter includes but looks beyond the sources that are normally reviewed in creating statements on graduate attributes. The review was part of the work done in developing the taxonomy of engineering competencies. Given its somewhat unique genesis, context, and perspective, this particular taxonomy provides an interesting case study of how literature, theory, and research-based evidence can be combined to form statements of graduate attributes for a specific educational discipline.

## Introduction

The general impetus which motivated the development of the taxonomy of engineering competencies described in this chapter was the societal change in South Africa after the demise of Apartheid. This change led to educational *massification* and the typical problems associated with it – under-prepared students, large classes, and a diverse first year intake all of which contributed to substantial attrition and academic failure.

In describing the development of the taxonomy of engineering competencies – hereafter referred to simply as the *taxonomy* – the chapter is divided into three parts. Part 1 begins with a brief review of the concepts of quality and curriculum responsiveness. This provides a theory-based position for identifying the stakeholders in engineering education and their concerns. Following this, attention is given to the

L.C. Woollacott (✉)
School of Chemical and Metallurgical Engineering,
Faculty of Engineering and Built Environment, University of the Witwatersrand,
Johannesburg, Private Bag 3, WITS 2050, South Africa
e-mail: lorenzo.woollacott@wits.ac.za

A.S. Patil and P.J. Gray (eds.), *Engineering Education Quality Assurance: A Global Perspective*,
DOI 10.1007/978-1-4419-0555-0_21, 

important issue of what is understood by the term *competency*. A review of the literature relating to engineering competencies constitutes Part 2 of the chapter. It includes listings of graduate competencies and attributes that are considered relevant and significant to an articulation of the goals of engineering education. The review is based on the literature search carried out during the development of the *taxonomy*. To bring the review up to date, literature and taxonomies that have emerged since the *taxonomy* was formulated in 2002 are also discussed in Part 2. Part 3 presents the *taxonomy* and describes its development as a case study that draws on the principles in Part 1 and the information gleaned from literature that is presented in Part 2.

## Part 1: Some Preliminaries – Quality and Competency

### *Identifying the Stakeholders in Engineering Education*

Quality is a complex trait. It includes not only a judgment of the extent to which a product or service meets a range of expectations, and is free of defects, but also how a customer experiences the product or service, both in part and as a whole (Sinha and Willborn 1985, p. 4). To define quality, therefore, one must identify the expectations of customers regarding the performance of the products or services they receive.

But, in the sphere of higher education, what do we mean by *customer*? To answer this question, it is helpful to begin with the concept of *curriculum responsiveness*. This is the idea that a curriculum (the educational program as a whole[1]) must be appropriately *responsive* to the legitimate expectations, requirements, and interests of stakeholders regarding how the program functions and what it delivers. Moll (2004), in synthesizing relevant theory, distinguishes between the following four kinds of curriculum responsiveness and, in doing so, identifies the four primary *stakeholders* in higher education.

1. *Economic responsiveness*. This has to do with how the curriculum "is responsive to the prevailing labor market by incorporating the necessary high level qualifications, knowledge and skills demanded by a modern, diversified economy" (p. 4). Here the *stakeholders* of engineering education are the economy and the labor market.
2. *Disciplinary responsiveness*. This has to do with how the curriculum "is responsive to the nature of its underlying discipline by ensuring a close coupling between the way in which knowledge is produced and the way students are educated in

[1] "*Curriculum* comprises all the opportunities for learning provided by an educational institution. These include the formal program of lessons in the timetable and the climate of relationships, attitudes and styles of behavior promoted within the institution as a whole" (Department of Education and Science for England and Wales, 1980, in Simelane 2006, p.32).

the discipline area" (p. 5). Here the *stakeholder* is the discipline – engineering in general and/or a particular branch of engineering.

3. *Cultural/Societal responsiveness*. This has to do with how the curriculum "is responsive to the cultural diversity of students and society by incorporating multiple cultural reference points that acknowledge diversity and constitute various alternative learning pathways for students" (p. 7). Here the *stakeholder* is society at large.
4. *Learner responsiveness*. This has to do with how the curriculum "is responsive to the learning needs of students by teaching them in terms that are accessible to them and assessing them in ways that they can understand" (p. 8). Here the *stakeholder* is the student.

## Responsiveness: The Provision of Quality Educational Programs

Accreditation standards used by professional engineering bodies relate directly to economic and disciplinary responsiveness: standards are used with the intention of making sure that graduates from accredited programs have the knowledge, skills, and dispositions (values/attitudes/commitments) demanded by the labor market and are competent to participate in and contribute as professionals to the practice of a particular branch of engineering.

In regard to the nature of societal and learner responsiveness, the South African context provides interesting examples. After the demise of Apartheid, considerable political transformation has taken place in which the issue of education has been key. A particularly pressing problem was how to restructure educational systems so that they address the very significant shift that occurred in the demographics and educational backgrounds of entrants to higher education. Learner responsiveness was a major concern here because of the very high levels of student *under-preparedness* for higher education programs (Pinto 2001; Woollacott et al. 2003). In response to this concern, a national policy was created to guide the South African educational restructuring effort.

The following list is an extract from a bulletin of the South African Qualifications Authority (SAQA) (South African Qualifications Authority 1997, p. 8). The extract spells out the general, nontechnical or core competencies – termed critical cross-field outcomes – which any educational program in South Africa is required to develop in learners. The last item in the list expresses very clearly the concern that an educational program should facilitate both professional and personal development since both the provision of suitably qualified professionals and the personal change attained through their educational experience have a positive impact on and enrich society. The Minster of Education put it this way, an educational program should facilitate the development in graduates of "intellectual capabilities and skills that can both enrich society and empower themselves and enhance economic and social development" graduates should be able to: (Department of Education 2007, p. 3).

1. Identify and solve problems in which responses display that responsible decisions using critical and creative thinking have been made.
2. Work effectively with others as a member of a team, group, organisation or community.
3. Organise and manage oneself and one's activities responsibly and effectively.
4. Collect, analyze, organise and critically evaluate information.
5. Communicate effectively using visual, mathematical and/or language skills in the modes of oral and/or written presentation.
6. Use science and technology effectively and critically, showing responsibility towards the environment and health of others.
7. Demonstrate an understanding of the world as a set of related systems by recognising that problem-solving contexts do not exist in isolation.
8. To contribute to the full personal development of each learner and the social and economic development of society at large, it must be the intention underlying any program of learning to make an individual aware of the importance of:

- Reflecting on and exploring a variety of strategies to learn more effectively
- Participating as responsible citizens in the life of local, national and global communities
- Being culturally and esthetically sensitive across a range of social contexts
- Exploring education and career opportunities
- Developing entrepreneurship

Cultural/societal, economic, and disciplinary responsiveness are made more explicit in a second extract from South African government policy documents (South African Qualifications Authority 2000, p. 14) which states that an educational program should:

- provide benefits to society and the economy through enhancing citizenship, increasing social and economic productivity, providing specifically skilled/professional people and transforming and redressing legacies of inequity;
- add value to qualifying learners in terms of enrichment of the person through the provision of status, recognition, credentials, and licensing, marketability and employability; and the opening-up of access routes to additional education and training.

These extracts imply that educational programs should aim to satisfy the legitimate expectations of all four groups of stakeholders simultaneously.

## *Competency and Graduate Attributes*

In simple terms, *competence* means "having the necessary skill or knowledge to do something successfully" and comes from the Latin *competere* "to be fit or proper" (Compact Oxford English Dictionary on AskOxford.com). As applied to professionals such as engineers it conveys the idea of possessing sufficiently the capability, skill, aptitude, proficiency, and expertise required to perform professional duties effectively. A more rigorous definition sees competency as "an underlying characteristic of an individual that is causally related to (causes or predicts) criterion-referenced effective and/or superior performance in a job or situation" (Spencer and Spencer 1993, p. 9). It is important to recognize that the criteria used to assess the level of competence are closely linked to the characteristic of the product or service to be provided, that is, the intended consequences of the task(s) that are performed. This link is brought out very clearly in the definition of competency that sees it as the ability to produce intended consequences without creating unintended consequences (Argyris and Schon 1974, pp. 6, 29). Passow (2007, p. 1) pulls these ideas together well in her definition of *competencies* as:

> the knowledge, skills, abilities, attitudes, and other characteristics that enable a person to perform skillfully (i.e., to make sound decisions and take effective action) in complex and uncertain situations such as *professional work* [emphasis added], civic engagement, and personal life.

The above definitions draw attention to three basic elements of the concept of competency.

- It is a latent, acquired, or developed attribute (an ability, capacity, or characteristic) possessed by a person.
- It is related to the intentional execution of tasks.
- It implies a value judgment on the quality of the ability, capacity, or characteristic and that this quality is assessed against formally or informally defined criteria by observing or measuring how effectively intended tasks are performed.

It is important to emphasize that competency and performance are linked. Competencies are internal attributes while performance is the result of these attributes in action. The quality of a competency is assessed by measuring the quality of the relevant performance. There is, however, some ambiguity in the literature about the meaning of *performance* in regard to task or work performance. As Williams (2002, chapters 4 and 5) explains, two positions exist. The first sees performance as output and assesses its quality in terms of deliverables and the bottom line – sales made, units manufactured, defects found, etc. Equivalent measures of performance in an educational environment would be grades achieved. The second position sees performance more in terms of the activity that lies behind output. In this case, the focus is on the behaviors required for such activity to be productive and the quality of performance is assessed in terms of measurable behavioral criteria. For example, one aspect of work performance is the ability and disposition to innovate. Performance as behavior would ask whether a person demonstrates innovative behaviors such as "does not do new things"; "does things to improve performance that are new to the job or work unit, new to the organisation, new to the industry" or are so new they "transform an industry" (Spencer and Spencer 1993, p. 27). In contrast, performance as output would ask how many identifiable innovations have been *delivered*.

Our discussion of the term *competency* emphasizes the mandate of engineering education to develop in students those attributes that a graduate engineer must possess to be capable of (1) producing desired engineering outcomes efficiently, and (2) acting in a manner that is productive and consistent with professional standards. By focusing on the importance of the quality of productive activity, it expands the range of educator attention beyond knowledge and skills to include affective and behavioral issues.

## A Generic Classification of the Elements of Competency

Campbell et al. (1993), working in the area of industrial psychology and human resource management, developed a model of the generic determinants of competency that they claimed was comprehensive in scope. The claim is well supported (Williams 2002, p. 99). The Campbell et al. (1993) model is presented as Table 1 with only minor modifications to its language.

The model recognizes three categories of attributes. The first – declarative knowledge – has to do with knowledge that can be communicated. The second has to do with skills and the knowledge intimately associated with skills – procedural knowledge. This kind of knowledge cannot be communicated as it is acquired through practice and the experience of becoming proficient in the associated skill. Subcategories of each kind of knowledge are listed in Table 1

**Table 1** The generic elements of competency (Adapted from Campbell et al. 1993, and reproduced here with the kind permission of John Wiley & Sons, Inc.)

| Attributes | Subcategories | Factors which influence the quality of the attributes |
|---|---|---|
| Declarative knowledge | Facts<br>Principles<br>Goals<br>Self-knowledge | (1) *Aptitudes (and values*[a]*)*: ability, personality, interests<br>(2) *Prior learning experience*: education, training, experience<br>(3) Interactions between aptitudes (values[a]) and prior learning experience |
| Procedural knowledge and skill | Cognitive skill<br>Psychomotor skill<br>Physical skill<br>Self-management skill<br>Interpersonal skill | (1) *Aptitudes (and values*[a]*)*: ability, personality, interests<br>(2) *Prior learning experience*: education, training, practice, experience<br>(3) Interactions between aptitudes (values[a]) and prior learning experience |
| Motivation (dispositions[a]) | Choices about:-<br>(a) whether to perform<br>(b) the level of effort<br>(c) the degree of persistence | Depends on which motivation theory is used |

[a]Added by this author

Knowledge has been classified in other ways but these generally fit with the categories and subcategories used in the model. For example, in her definition of competencies, Passow (2007) refers to the four kinds of knowledge that Anderson et al. (2001) include in their taxonomy of knowledge. These are factual knowledge (terminology and details), conceptual knowledge (classifications, principles, theories, and models), procedural knowledge (knowing how and when to use specific skills and methods), and meta-cognitive knowledge (self-knowledge and both how and when to use cognitive strategies for learning and problem-solving).

The third category in Campbell's model is *motivations*. This has been expanded in the table to include *dispositions*. The reason for this elaboration is that the notion of *dispositions* incorporates a wider range of affective traits, attributes, and commitments along with motivation. It draws attention to how all these factors can influence the way a person actually marshals knowledge and skills and brings them to bear in the performance of his/her work.

## Part 2: Perspectives on Engineering Competencies from the Literature

Various perspectives on engineering competency are found in the literature and are discussed in the sections that follow. The progression of the following discussion is similar to that followed in the formulation of the taxonomy. It starts with accreditation standards that describe the competencies that engineering graduates should possess and moves progressively through literature where the focus is more on generic competencies associated with the effective performance of work in general. These are presented in various tables which were primary sources from which the taxonomy was derived. Examples of statements relating to relevant competencies that have emerged since the taxonomy was first formulated in 2002 are also discussed and, in some cases, are also presented in tables.

### *Perspectives from Accreditation Standards*

The literature review behind the taxonomy looked at statements of required learning outcomes found in documents published by national bodies responsible for the accreditation of engineering programs in the USA, South Africa, Australia, Canada, New Zealand, and the UK. Table 2 summarizes and compares the first two of these and shows, not surprisingly, a high degree of consensus. The examination of documentation from the other accrediting bodies mentioned shows a similar

**Table 2** Summaries and comparison of engineering education accreditation standards in the United States and South Africa (reproduced here with the kind permission of ABET Inc. and ECSA)

| Accreditation Board for Engineering and Technology (ABET Inc.) (2007) (United States) | Engineering Council of South Africa (ECSA) (2004) |
|---|---|
| Engineering programs must demonstrate that their students attain the following outcomes: | A graduate must be competent to ... |
| (a) Apply knowledge of mathematics, science, and engineering | Apply knowledge of mathematics, basic science, and engineering sciences ... to solve engineering problems |
| (b) Design/conduct experiments and analyze and interpret data | Design and conduct investigations and experiments |
| (c) Design a system, component or process to meet desired needs within realistic constraints ... | Perform creative, *procedural* and *nonprocedural* design and synthesis of components, systems, engineering works, products, or processes |
| (d) Function on multidisciplinary teams | Work effectively as an individual, in teams and multidisciplinary environments |
| (e) Identify, formulate, and solve engineering problems | Identify, assess, formulate, and solve *convergent* and *divergent* engineering problems creatively and innovatively |
| (f) Understanding of professional and ethical responsibility | Demonstrate critical awareness of the need to act professionally and ethically and exercise judgment and take responsibility within own limits of competence |
| (g) Communicate effectively | Communicate effectively, both orally in writing and, with engineering audiences and the community at large |
| (h) Broad education necessary to understand the impact of engineering solutions in a global/social context | Demonstrate *critical awareness* of the impact of engineering activity on the social, industrial, and physical environment |
| (i) Recognition of the need for and the ability to engage in life-long learning | Engage in independent learning through well-developed learning skills |
| (j) Knowledge of contemporary issues | ---- |
| (k) Use the techniques, skills, and tools needed for engineering practice | Use appropriate engineering methods, *skills*, and tools including those based on information technology |

degree of consistency. Many of these accreditation standards have been updated since 2002 and the reader is referred to the relevant Web sites for these. (A list of these sites is appended to the references at the end of the chapter.)

The International Engineering Alliance (IEA) published an important article on the desired attributes of engineering graduates (International Engineering Alliance 2005). The IEA is a forum for six international accreditation accords including the Washington, Sydney and Dublin Accords (see http://www.ieagreements.com). These accords are concerned with the globalization of accreditation standards through a process of mutual recognition of the national standards of the signatories to the accords. The article provides a benchmark for the mutual recognition process and the relevant content is presented here as Table 3.

In the UK, work in the EPC (Engineering Professor's Council) produced a statement about outcome standards for engineering programs that was published in an article by Maillardet (2004). The statement resulted from work toward a national accreditation standard. It used the design process as the basis for framing the statement of required graduate competencies. The statement has a somewhat different format and wording than other accreditation standards and so is shown here as a separate table (Table 4).

**Table 3** The IEM graduate attributes profile (Extracted from Graduate Attributes and Professional Competencies, International Engineering Alliance 2005, and reproduced here with the kind permission of the IEA Secretariat)

| Topic | Graduate attribute |
|---|---|
| 2. Knowledge of engineering sciences | Apply knowledge of mathematics, science, engineering fundamentals, and an engineering specialization to the conceptualization of engineering models |
| 3. Problem analysis | Identify, formulate, research literature, and solve complex engineering problems reaching substantiated conclusions using first principles of mathematics and engineering sciences |
| 4. Design/ Development of solutions | Design solutions for complex engineering problems and design systems, components or processes that meet specified needs with appropriate consideration for public health and safety, cultural, societal, and environmental considerations |
| 5. Investigation | Conduct investigations of complex problems including design of experiments, analysis and interpretation of data, and synthesis of information to provide valid conclusions |
| 6. Modern tool usage | Create, select, and apply appropriate techniques, resources, and modern engineering tools, including prediction and modeling, to complex engineering activities, with an understanding of the limitations |
| 7. Individual and team work | Function effectively as an individual, and as a member or leader in diverse teams and in multidisciplinary settings |
| 8. Communication | Communicate effectively on complex engineering activities with the engineering community and with society at large, such as being able to comprehend and write effective reports and design documentation, make effective presentations, and give and receive clear instructions |
| 9. The engineer and society | Demonstrate understanding of the societal, health, safety, legal, and cultural issues and the consequent responsibilities relevant to engineering practice |
| 10. Ethics | Understand and commit to professional ethics and responsibilities and norms of engineering practice |
| 11. Environment and sustainability | Understand the impact of engineering solutions in a societal context and demonstrate knowledge of and need for sustainable development |
| 12. Project management and finance | Demonstrate a knowledge and understanding of management and business practices, such as risk and change management, and understand their limitations |

(continued)

**Table 3** (continued)

| Topic | Graduate attribute |
|---|---|
| 13. Life long learning | Recognize the need for, and have the ability to engage in independent and life-long learning |

Notes:

(1) Item 1 in the IEM table is not relevant as it refers to a type of educational institution and, therefore, it was omitted from Table 3.

(2) The IEM profiles for technologists and technicians have not been included in this table.

(3) Complex engineering problems and complex activities as used in the IEM Profile are as follows:

*Complex Engineering Problems* are those which cannot be resolved without in-depth engineering knowledge and having some or all of the following characteristics:

- Involve wide-ranging or conflicting technical, engineering and other issues
- Have no obvious solution and require abstract thinking, originality in analysis to formulate suitable models
- Requires in-depth knowledge that allows a fundamentals-based first principles analytical approach
- Involve infrequently encountered issues
- Are outside problems encompassed by standards and codes of practice for professional engineering
- Involve diverse groups of stakeholders with widely varying needs
- Have significant consequences in a range of contexts
- Are high level problems possibly including many component parts or subproblems

*Complex Engineering Activities* are those that have some or all of the following characteristics:

- Involve the use of diverse resources (and for this purpose resources include people, money, equipment, materials, information, and technologies)
- Require resolution of significant problems arising from interactions between wide-ranging or conflicting technical, engineering or other issues,
- Involve creative use of knowledge of engineering principles in novel ways.
- Have significant consequences in a range of contexts
- Can extend beyond previous experiences by applying principles-based approaches

**Table 4** The EPC outcome standards (Extracted from Maillardet 2004, pp. 33–55, and reproduced here with the kind permissions of Taylor & Francis Books UK)

| Primary elements | Elaboration |
|---|---|
| 1. Ability to exercise key skills in the completion of engineering-related tasks | The key skills for engineering are communication, information technology, application of number, working with others, problem-solving, improving own learning, and performance. |
| | Ability to ... |
| 2. Ability to transform existing systems into conceptual models | Elicit and clarify client's true needs<br>Identify, classify, and describe engineering systems<br>Define real target systems in terms of objective functions, performance specifications, and other constraints (i.e., define the problem).<br>Take account of risk assessment, and social and environmental impacts, in the setting of constraints (including legal, health, and safety issues).<br>Select, review, and experiment with existing engineering systems to obtain a database of knowledge and understanding that will contribute to the creation of specific real target systems.<br>Resolve difficulties created by imperfect and incomplete information.<br>Derive conceptual models of real target systems, identifying the key parameters. |

(continued)

**Table 4** (continued)

| Primary elements | Elaboration |
|---|---|
| 3. Ability to transform conceptual models into determinable models | Construct determinable models over a range of complexity to suit a range of conceptual models<br>Use mathematics and computing skills to create determinable models by deriving appropriate constitutive equations and specifying appropriate boundary conditions<br>Use industry standard software tools and platforms to set up determinable models<br>Recognize the value of models of different complexity and limitations of their application |
| 4. Ability to use determinable models to obtain system specifications in terms of parametric values | Use mathematics and computing skills to manipulate and solve determinable models; and use data sheets in an appropriate way to supplement solutions.<br>Use industry standard software platforms and tools to solve determinable models<br>Carry out a parametric sensitivity analysis<br>Critically assess results and, if inadequate or invalid, improve knowledge database by further reference to existing systems, and/or performance or determinable models |
| 5. Ability to select optimum specifications and create physical models | Use objective functions and constraints to identify optimum specifications<br>Plan physical modeling studies based on determinable modeling, to produce critical information<br>Test and collate results feeding these back into determinable models |
| 6. Ability to apply the results from physical models to create real target systems | Write sufficiently detailed specifications of real target systems, including risk assessments and impact statements<br>Select production methods and write method statements<br>Implement production and deliver products fit for purpose, in a timely and efficient manner<br>Operate within relevant legislative frameworks |
| 7. Ability to critically review real target systems and personal performance | Test and evaluate real systems in service against specification and clients needs<br>Recognize and make critical judgments about related environmental, social, ethical, and professional issues<br>Identify professional, technical, and personal development needs and undertake appropriate training and independent research |

## *The CDIO Perspective*

CDIO (Conceive, Design, Implement and Operate) is a multinational reform initiative that is concerned to close the gap between engineering education and engineering practice while remaining faithful to both engineering professionalism and the need "to provide quality education in technical fundamentals" (Crawley 2002). The gap between engineering education and practice is explained as the result of a shift that occurred in the middle of the last century in the way that engineering was taught (Crawley

2002; Grimson 2002). The shift was characterized by the increasing prominence given to engineering science in engineering education as compared with the more traditional emphasis on practical engineering (Grimson 2002).

In an effort to close this gap, the CDIO initiative reevaluated the goals of engineering education from the perspective of modern engineering practice and developed a generic syllabus (the CDIO Syllabus) that used design (or, more accurately, CDIO) as its chief organizing principle. As a statement of the goals of engineering education, the CDIO Syllabus became the foundation for the curriculum redesign component of the reform initiative (Crawley 2002; Crawley et al. 2007). It was developed as a collaborative effort between a range of engineering schools (aerospace, mechanical, and electronics engineering) at MIT and three Swedish universities over a 3-year period based on work involving focus groups, surveys, workshops, and peer reviews (Crawley 2002).

The CDIO Syllabus details the many, interrelated processes, knowledge, skills, and attributes involved in engineering a technical system or product from its conception, through design, construction, and implementation, through its operation and eventual life-end and disposal. It also details the external, societal, enterprise, and business contexts in which such engineering is conducted and the personal, interpersonal, and professional skills needed for competent performance of the relevant engineering tasks and processes. The syllabus constitutes the most detailed statement on required graduate competencies currently found in the literature (Woollacott 2007). An abbreviated version and discussion of the CDIO syllabus appear in chapter "CDIO and Quality Assurance: Using the Standards for Continuous Programme Improvement" by Brodeur and Crawley and the full version may be found in Crawley et al. (2007, pp. 257–268) or on the CDIO website (http://www.cdio.org).

## *Perspectives from Surveys of Engineering Employers and Practicing Engineers*

Over the years, many surveys have been conducted to determine which competencies engineering employers look for in engineering graduates (Boeing 1966; Young 1986; Natriello 1989; Busse 1992; Augustine 1994; Kemp 1999; Skakoon and King 2001; de Jager and Nieuwenhuis 2002; World Chemical Engineering Council 2004; Crawley et al. 2007, pp. 58–59). For example, the top five personal qualities/skills employers seek, according to the National Association of Colleges and Employers (2008) *Job Outlook 2009* survey, are:

1. Communication skills (verbal and written)
2. Strong work ethic
3. Teamwork skills (works well with others)
4. Initiative
5. Analytical skills

In his book on studying engineering, Landis (2007, p. 21) lists the top six factors to which US employers refer, in his experience, when considering a graduate engineer for employment. They are as follows:

- Personal qualifications – including maturity, initiative, enthusiasm, poise, appearance, and the ability to work with people.
- Scholastic qualifications – as shown by grades in all subjects or in a major field of study.
- Specialized courses students have taken in particular fields of work.
- Ability to communicate effectively, both orally and in writing.
- Kind and amount of employment while at college.
- Experience in campus activities, especially participation and leadership in extra-curricula life.

A South African study by de Lange (2000) concentrated on eliciting from employers their opinions about the nontechnical attributes they looked for in graduates. *Nontechnical* competencies that de Lange identified as being potentially relevant were grouped into appropriate clusters. Table 5 presents the results of the survey organized by the clusters and the associated competencies that formed the basis of the survey questionnaire used in the study.

An in-depth study of the competencies engineering employers and practicing engineers considered important was conducted recently by Passow (2007). From a comprehensive literature review, she identified 12 studies that had been carried out from 1992 to 2007 (National Society of Professional Engineers 1992; Turley 1992; Evans et al. 1993; American Society of Mechanical Engineers 1995; Benefield et al. 1997; Shea 1997; Koen and Kohli 1998; Lang et al. 1999; Bankel et al. 2003; Saunders-Smith 2005; Lattuca et al. 2006). Of these, ten asked respondents to rate desired graduate competencies on a five-point scale. Passow (2007) reexamined the data in the ten studies using a meta-analysis methodology to obtain a synthesized opinion from the 5,978 respondents to the 19 surveys covered in these ten studies. Passow's (2007) paper also includes 12 tables that summarize the wording used to describe the various competencies included in the 19 surveys.

Passow's (2007) analysis involved mapping the competencies onto the 11 ABET competencies ((a)–(k), see Table 2), transforming the data to a common metric, and using multiple comparison procedures and a careful statistical analysis to distinguish the relative importance assigned by respondents to the different sets of competencies. Relative importance was reported on a five-point scale ranging from +2.5 to –2.5 where 0 represented the *ABET mean* – the average rating for all the competencies that mapped onto the 11 ABET competencies. Competencies that did not map onto the ABET competencies were analyzed separately.

Passow's (2007) findings are summarized in Table 6. Among the ABET competencies, six levels of perceived importance were identified by determining which ratings were statistically different and which were not. As indicated in Table 6, eight competency sets that did not map onto the ABET categories were also shown to fall into or between these six levels of perceived importance. Passow (2007) makes an interesting distinction between competencies and *bodies of knowledge*

**Table 5** Nontechnical skills important for engineering graduates (Extracted from de Lange 2000, and reproduced here with the kind permission of G. de Lange)

| | Ranked in order of perceived importance | Other skills in questionnaire[a] |
|---|---|---|
| *Functional skills* | The basic skills applied to tasks such as speaking, reading, and writing. They form part of larger actions such as instructing and leading a team of workers. | |
| (1) Communication | The ability to exchange, transmit, and express knowledge and ideas to achieve set objectives. | |
| (IR = 1)[b] (RI = 98%) | Verbal communication, listening, explanation, technical report writing, reading, visual and graphic presentation, demonstration | Teaching, grievance handling, conversation, negotiation, conflict management, visual presentation, meeting procedure, interviewing, presentation, selling, persuasion, instruction |
| (2) Information management | The ability to arrange, sort, retrieve data, knowledge, and ideas. | |
| (IR = 7) (RI = 84%) | Logical thinking, analysis, prioritizing, reporting, computer application, recording, collection | Retrieval, research, organization, scheduling, synthesizing, sorting, valuation |
| (3) Creative thinking and problem solving | The ability to solve existing and anticipated problems through creative innovative and analytical means. | |
| (IR = 2) (RI = 96%) | Problem analysis, observing, questioning, interpreting, investigating, innovating, anticipating, formulating | Forecasting, being creative, interpretation, conceptualisation, prediction, facilitation |
| *Adaptive skills* | Skills required to "fit in" and contribute as a valuable member in the work place. | |
| (4) Personal style and self-management | Indicators of general outlook, personal appearance, values, goals, and motivation. | |
| (IR = 5) (RI = 87%) | Is motivated, is responsible, is self-confident, is honest, has integrity, is disciplined, is enthusiastic, has positive self-esteem, is adaptable, is determined, is flexible, is conscientious, is ethical, is dependable, is stable | Is assertive, is persistent, is sincere, is patient, is mature, has good appearance, is objective |

| | | |
|---|---|---|
| (5) Work-related dispositions and attitudes (IR=4) (RI=91%) | Indicators of personal work orientation, work values, attitudes, and understanding of the work environment. | |
| | Thoroughness, willing to learn and be trained, committed to job, interest, pride in work, respect for property, understands teamwork, precise, makes extra effort, task orientated, punctual, good work habits, takes initiative, understands work environment, handles pressure and stress | Team member, willing to be trained, accepts criticism, gives credit, open-minded, pride in work, respectful, self-control, takes risks |
| (6) Group effectiveness and teamwork (IR=3) (RI=92%) | The ability to use the correct combination of interpersonal skills to direct and guide a team to complete tasks and attain goals. | |
| | Cooperates, is responsive, is helpful, coordinates, is compatible, has group process skills, is tactful, is even tempered, is sensitive to cultural diversity, leads and manages, recruits ideas, summarizes | Puts people at ease, negotiates, solicits, has social commitment, is hospitable, is outgoing, supervises, praises, counsels, has empathy, is persuasive |
| (7) Organizational Effectiveness and teamwork (IR=5) (RI=85%) | The ability to effectively contribute toward the successful completion of a set of organizational goals. | |
| | Meets deadlines, works to schedule, is goal orientated, assumes responsibility, puts theory into practice, works under pressure, prioritizes, makes suggestions, sets objectives, manages time, handles stress, follows procedures, motivates, co-ordinates | Is goal directed, has vision, delegates, leads, directs, administers, manages, supervises, instructs, applies policies, recommends |

[a]Details obtained from the author through private correspondence.
[b]Results of the survey are indicated by Importance Ranking (IR) and % Relative Importance (RI).

**Table 6** Results of a meta-analysis of the opinions of employers and practicing engineers in regard to desired graduate competencies (Extracted from Passow 2007, and reproduced here with the kind permission of H. J. Passow)

| Importance level | ABET criterion | Abbreviation of ABET competency set (importance rating[a]) | Non-ABET-based competency sets[b] | |
|---|---|---|---|---|
| | | | Skill-related | Attitude-related |
| 1 | e | Problem-solving (1.02) | | |
| | g | Communication (0.9) | | |
| 1.5 | | | Decision making | Commitment to achieving goals |
| 2 | f | Ethics (0.61) | | |
| 3 | i | Life-long learning (0.39) | | |
| | | | Able to transition to the industrial environment | Mature, responsible, and open-minded with a positive attitude toward life |
| | | | Project management | Personal skills and attributes[c] |
| | | | Leadership skills | |
| 4 | b | Ability to conduct and evaluate experiments (0.2) | | |
| | d | Team-work skills (0.18) | | |
| | k | Ability to use engineering tools (0.17) | | |
| | c | Design skills (0.06) | | |
| 5 | a | Maths, science, engineering knowledge (–0.05) | | |
| 5.5 | | | Business skills | |
| 6 | j | Knowledge of contemporary issues (–1.39) | | |
| | h | Knowledge to assess impact (–1.54) | | |

[a]The importance rating is a relative indication of importance on a scale from +2.5 to –2.5 where 0 is the ABET mean.
[b]The importance ratings for non-ABET-mapped competencies are not included in the table.
[c]These include initiative and willingness to take risks, perseverance and flexibility, creative and critical thinking, curiosity, life-long learning, time and resource management, and awareness of one's personal knowledge, skills and attitudes

and noted that competencies were uniformly rated by practicing engineers as being more important (levels 1 to 4) than bodies of knowledge (levels 5 and 6) – business skills being the only exception (level 5.5).

## *Perspectives from Human Resource Management Literature*

The perspectives described in the previous section were based directly or indirectly on the results from workplace surveys. A different method for soliciting information from the work place has been used for over 20 years by the McBer Consulting Agency. Their methods and findings have been published in a book entitled *Competency at Work: Models for Superior Performance* (Spencer and Spencer 1993). The work is widely respected (Williams 2002, pp. 102–114).

The motivation for the Agency's work was the need to select personnel and to objectively distinguish between ordinary performers and superior performers. Their approach was to develop a competency model for a particular job by identifying superior performers in that job, interviewing them to discover behavioral traits that characterized their work performance and comparing these findings with those from interviews of "ordinary" performers.

The interviews were conducted by experienced human resource investigators trained in a formalized methodology that had been developed by the Agency over the years. Their task was to identify characteristic behaviors of superior performers and to describe each one in the form of a short narrative description along with measurable behavioral indicators. For example, they identified eight behavioral indicators relating to *self control*. These were: losses control, avoids stress, resists temptations, controls emotions, responds calmly, manages stress effectively, responds constructively, and calms others. Once the set of distinguishing competencies and the related behavioral indicators had been identified, they were arranged into relevant clusters of competencies, which then formed the competency model for the particular job.

Over a span of 20 years, more than 100 trained investigators have developed 286 competency models in over 20 countries. The models cover technical/professional job types as well as jobs in the fields of human service, entrepreneurship, sales/marketing/trading, and managers (in industry, government, military, health care, education, and religious organizations). Technical professionals or *knowledge workers* are defined as "individual contributors whose work involves the use of technical (as opposed to human services) knowledge" (Spencer and Spencer 1993, pp. 161–163). Models for technical professionals have been developed for software developers, engineers, and applied research scientists.

Drawing on this breadth of experience, the Agency extracted generic competencies and behavioral indicators from the models and arranged them into a *competency dictionary*. The dictionary consists of 6 clusters of distinguishing competencies, 21 groups of competencies, and, depending on how you count them, 35 or 28 generic competencies with 360 or 278 behavioral indicators. The dictionary is summarized in Table 7.

**Table 7** A summary of the McBer competency dictionary (Extracted from Spencer and Spencer 1993, chapters 4 to 9, and reproduced here with the kind permission of John Wiley & Sons, Inc.)

| Distinguishing competency cluster | Competency group | Competency | Number of behavioral indicators |
|---|---|---|---|
| (1) Achievement and action | Achievement orientation | Intensity and completeness of achievement orientation | 9 |
| | | Achievement impact | 7 |
| | | Degree of innovation | 5 |
| | Concern for order, quality, accuracy | Concern for order, quality, and accuracy | 9 |
| | Initiative | Time dimension. | 11 |
| | | Self-motivation, amount of discretionary effort. | 8 |
| | Information seeking | Information seeking | 8 |
| (2) Helping and human service | Interpersonal understanding | Depth of understanding of others | 7 |
| | | Listening and responding to others | 7 |
| | Customer service orientation | Focus on client's needs | 13 |
| | | Initiative (discretionary effort) to help or serve others | 7 |
| (3) Impact and influence | Impact and influence | Actions taken to influence others | 10 |
| | | Breadth of influence, understanding or network | 9 |
| | Organizational awareness | Depth of understanding of organization | 8 |
| | Relationship building | Closeness of relationships built | 9 |
| (4) Managerial | Developing others | Intensity of developmental orientation and completeness of developmental action | 11 |
| | | Number and rank of people developed or directed | 9 |
| | Directiveness: Assertiveness and use of positional power | Intensity of directiveness | 11 |
| | Teamwork and cooperation | Intensity of fostering teamwork | 9 |
| | | Size of team involved | 6 |
| | | Amount of effort or initiative to foster teamwork | 6 |
| | Team leadership | Strength of leadership role | 9 |
| (5) Cognitive | Analytical thinking | Complexity of analysis | 7 |
| | | Size of problem addressed | 5 |
| | Conceptual thinking | Complexity and originality of concepts | 8 |
| | Technical, professional, managerial expertise | Depth of knowledge | 8 |
| | | Breadth of managerial experience | 7 |
| | | Acquisition of expertise | 5 |
| | | Distribution of expertise | 7 |

(continued)

**Table 7** (continued)

| Distinguishing competency cluster | Competency group | Competency | Number of behavioral indicators |
|---|---|---|---|
| (6) Personal effectiveness | Self-control | Self-control | 8 |
| | Self-confidence | Self-assurance | 8 |
| | | Dealing with failure | 6 |
| | Flexibility | Breadth of change | 8 |
| | | Speed of change | 5 |
| | Organizational commitment | Organizational commitment | 8 |
| | Other personal characteristics and competencies | Occupational preference, accurate self-assessment, affiliative interest, writing skills, visioning, upward communications, concrete style of learning and communicating, low fear of rejection, thoroughness | |

The generic categories in the dictionary cover from 80 to 98% of the specific categories found in the original competency models. On this basis, the Agency defined a generalized competency model for each of the five different job types mentioned above. It claims that each generalized model describes all jobs of each type in general but none in particular. Their competency model for technical professionals – including engineers – is presented in Table 8. It must be noted that the motivation behind the model is the identification of superior performers and this must be taken into account when using the dictionary. Its scope goes beyond the identification of graduate attributes to be used for accreditation or Quality Assurance purposes: in this regard the model should be taken only as describing *advanced attributes* that are desirable to find in engineering graduates, but are not necessarily expected in all graduates.

## *Perspectives on Work*

An engineer is first of all a worker and so competencies associated with effective work and productive work performance are relevant attributes to be expected in graduate engineers. Landis (2007, p. 84) identified ten different generic settings in which engineers may work (Table 9). The brief descriptions given in that table provide a view on engineering work that complements the other perspectives on engineering competencies described in this review. In the formulation of the taxonomy, two additional types of engineering work were added to Landis' list – maintenance work and entrepreneurial work.

Table 10 presents an augmented version of a taxonomy developed by Campbell et al. (1993) that claims to encompass the major performance components required in any kind of job. Williams (2002), in his review of the related literature, suggests

**Table 8** Summary of McBer's generalized competency model for technical professionals (Extracted from Spencer and Spencer 1993, p. 163, and reproduced here with the kind permission of John Wiley & Sons, Inc.)

| Competency | Relative weight[a] | Behavioral indicators |
|---|---|---|
| (1) Achievement orientation | 6 | Measures performance<br>Improves outcomes<br>Sets challenging goals<br>Innovates |
| (2) Impact and influence | 5 | Uses direct persuasion, facts, and figures<br>Gives presentations tailored to audience<br>Shows concern with professional reputation |
| (3) Conceptual thinking | 4 | Recognizes key actions, underlying problems<br>Makes connections and patterns |
| (4) Analytical thinking | 4 | Anticipates obstacles<br>Breaks problem apart systematically<br>Makes logical conclusions<br>Sees consequences, implications |
| (5) Initiative | 4 | Persists in problem solving<br>Addresses problems before asked to |
| (6) Self-Confidence | 3 | Expresses confidence in own judgment<br>Seeks challenges and independence |
| (7) Interpersonal understanding | 3 | Understands attitudes, interests, needs of others |
| (8) Concern for order | 2 | Seeks clarity of roles and information<br>Checks quality of work and information<br>Keeps records |
| (9) Information seeking | 2 | Contacts many different sources<br>Reads journals etc. |
| (10) Teamwork and cooperation | 2 | Brainstorms, solicits input<br>Credits others |
| (11) Expertise | 2 | Expands and uses technical knowledge<br>Enjoys technical work, shares expertise |
| (12) Customer service orientation | 1 | Discovers and meets underlying needs |

[a]The relative weight is the frequency with which the competency appeared in the specific competency models from which the generalized model was derived.

**Table 9** Descriptions of engineering work (Adapted from Landis 2007, pp. 84–87)

| Job function | Description |
|---|---|
| 1. Analysis | Does mathematical modeling of the physical and/or chemical aspects of problems using physics, chemical and engineering sciences, numerical and mathematical procedures, and engineering software. |
| 2. Design | Converts concepts and information into detailed plans and specifications for the development, manufacture or building of a product, component, system or process. |
| 3. Testing | Develops and conducts tests to verify that a selected design or product meets all specifications. |
| 4. Development | Develops products, processes or systems. Somewhere between the design and testing job functions. |
| 5. Selling | A technical liaison person between the company and the customer. Must be technically proficient to understand both the product and the customer's needs. |
| 6. Research | Involved in the search for new knowledge. Differs from a research scientist in that the motivation for the new knowledge is not knowledge for its own sake but knowledge that can be applied for the advancement of engineering practice. |

(continued)

**Table 9** (continued)

| Job function | Description |
|---|---|
| 7. Line management | Involved as technical staff in the supervision of designated aspects of the "production line" in engineering production enterprises. The involvement may be at various points in the supervision hierarchy from junior engineer to chief engineer to company president. |
| 8. Project management | Differs from line management in that personnel are organized according to a specific project and are responsible to ensure that the project is completed successfully, on time and within budget. |
| 9. Consulting | Provides "expert" technical services for a client on a contractual basis. |
| 10. Teaching | Works in an academic environment and is involved with teaching, research, and providing services in a specific area of an engineering discipline. |

**Table 10** A taxonomy of major performance components (Extracted from Campbell et al 1993, except for item 9, and reproduced here with the kind permission of John Wiley & Sons, Inc.)

| Performance component | Description |
|---|---|
| 1. Job-specific task proficiency | Proficiency in performing the core substantive or technical tasks that are central to the job. Job-specific performance behaviors that distinguish the substantive content of one job from another. |
| 2. Non-job-specific task proficiency | Proficiency in performing tasks or executing performance behaviors which are not specific to one's particular job – e.g., an engineer doing administration or sitting on the safety committee. |
| 3. Proficiency in written or oral communication | Proficiency in writing or speaking (independent of the correctness of the subject matter). |
| 4. Demonstrating effort | Consistent commitment to all job tasks, to working at a high level of intensity and the willingness to keep working under adverse circumstances and to expend extra effort when required. |
| 5. Maintaining personal discipline | The degree to which negative behaviors – such as alcohol abuse and absenteeism – are avoided. |
| 6. Facilitating peer and team performance | Supporting and helping peers and facilitating group functioning by being a good model, keeping the group goal directed, and reinforcing participation by other group members. |
| 7. Supervision and leadership | Influencing the performance of subordinates through interpersonal interaction and influence, modeling, goal setting, coaching, and providing reinforcement. Similar to (6) but supervisory leadership involves different performance determinants than peer leadership. |
| 8. Management and administration | Involves processes additional to those in (7) such as articulating goals for a production unit or enterprise, organizing people or resources to achieve these, monitoring progress, helping to solve problems or overcome crises that stand in the way of goal accomplishment, controlling expenditures, obtaining additional resources, and representing the unit in dealing with other units. |
| 9. Adaptive performance | "Ease of learning new tasks, confidence in approaching new tasks, flexibility and capacity to cope with change,"[a] "capacity to engage with new learning in coping with change,"[b] "developing oneself."[c] |

[a] (Hesketh and Neal 1999)
[b] (London and Mone 1999)
[c] (Williams 2002, p. 96)

that the taxonomy overlooks performances that have to do with self-development and adaptation to the fast pace of change characteristic of modern work environments (see also Hesketh and Neal 1999, and London and Mone 1999). Williams also noted terminology in the literature that differed from Campbell's as well as differences in emphasis and some differences in approach. On reflection, however, he concluded that (1) the differences were not very significant, (2) that Campbell's categories augmented with *adaptive performance* were an adequate general description of the major components of work performance, and (3) that the augmented taxonomy provides a reliable framework for making sure that no aspect of work performance is overlooked when analyzing the nature of any particular job.

Table 11 presents a perspective developed during the formulation of the taxonomy as a basic framework for describing the different aspects of an individual's work (Woollacott 2003). The rationale here is that different types of work functions require different profiles of competencies. For example, the competency mix needed for initiating work is different from the one needed for acquiring resources. The work functions in the taxonomy in the table are generic, however, in that each type of work function is associated with a similar competency profile in any context. For example, the initiation of a new project, a new task, a new procedure, or a new organization all involve similar kinds of functions although the extent and complexity of the competencies involved will be very different.

The perspective in Table 11 was formulated with inexperienced students in mind – students with limited experience or perception of what skills and attitudes are needed for satisfactory execution of tasks. The idea was to spell out to them what was involved and what they needed to give their attention to in order to develop the ability to execute work-related tasks in an ongoing and sustained way. It was considered to be particularly important for them to appreciate that besides the *core work functions* that get the job done, *support work functions* are very important to support, monitor, guide, and enable the efficient execution of core work functions.

The purpose of the taxonomy in Table 11 is to distinguish clearly what the two kinds of work functions involve. The first nine of these are self explanatory and are identified in various forms in other perspectives found in the literature. The tenth work function, *house keeping*, emphasizes the need to pay attention to resources – both one's own as well as those made available in the work environment. This work function is at the root of important factors such as tidiness, order, organizing resources effectively and caring properly for equipment, finances, and the capacity to sustain good work. This aspect of competency is considered to be of particular relevance to inexperienced learners, some of whom have little or no real awareness of the importance of these issues.

Table 12 presents the taxonomy of World of Work Skills developed by Evers et al. (1998). This taxonomy resulted from a project in Canada called *Make the Match* which was concerned with skills and human resource development, the relationship of education to work, and how to modify curricula to better prepare graduates for the world of work. The project was spear-headed by a nine-person task force (five corporate CEOs and four university presidents). Interestingly, it began with the intention of focusing on technical skills, but during the process of

**Table 11** A taxonomy of individual work functions (Woollacott 2003)

| Category | Work function |
|---|---|
| Line or core work functions | (1) *Initiating work*: This involves reflecting on and examining the technological, business, and social environments in which the organization operates. The purpose is to identify possible areas for expanding existing activities or markets and opportunities for new products, technology or services. |
| | (2) *Planning work*: The nature and requirements of the work must be properly analyzed and understood. The work to be done must be broken down into subtasks that are appropriate, manageable, well defined, and properly prioritized and scheduled. |
| | (3) *Acquiring resources for the work*: The staff, physical resources, information, and skills needed to accomplish the work must be acquired from the general market place or from within the organization, or from colleagues, consultants, suppliers or information systems. In some cases, this will involve learning by personal study or engaging in research and development. |
| | (4) *Performing subtasks*: The worker must bring his/her knowledge and the acquired resources to bear effectively to accomplish each of the required subtasks. |
| | (5) *Evaluating and synthesizing results*: The results from the different tasks must be brought together, properly evaluated and synthesized appropriately to achieve the overall objectives. |
| Support work functions | (6) *Managing the work*: This involves ensuring that the various aspects of work – both core and support work functions – are properly coordinated, are progressing in a satisfactory manner, and that problems that occur are dealt with appropriately. |
| | (7) *Evaluating effectiveness (productivity, profitability, quality, service and impact)*: This involves giving attention to the quality and effectiveness of the work effort and its results, and being sensitive to the impact the work makes or could make on the organization, the market, society, and the environment. This involves the examination of and the exercise of judgment about a broad range of factors from technical, financial, social, and legal, to the evaluation of alternative solutions, implementability, and issues of health and safety. |
| | (8) *Interacting with people*: The worker must be competent not only to work alone, but also as a member or leader of a team. Good teamwork involves making effective personal contributions, interacting with team members in ways that enhance their contributions, facilitating the productivity of the team and dealing effectively with interactional problems. In addition, a person may need to interact professionally with clients or with members of the public as a representative of the organization. |
| | (9) *Communication*: An important aspect of interacting with people is the ability to communicate effectively verbally, graphically and in writing with colleagues, clients, superiors, and subordinates. |
| | (10) "*Housekeeping*": This involves ensuring that the resources and capacity to do good work are maintained, sustained and, where necessary, are developed further. |

**Table 12** The taxonomy of world of work skills (Extracted from Evers et al. 1998, p. 40, and reproduced here with the kind permission of John Wiley & Sons, Inc.)

| | | |
|---|---|---|
| Managing self | | |
| Constantly developing practices and internalizing routines for maximizing one's ability to deal with the uncertainty of an ever-changing environment | Learning | Gaining knowledge from everyday experiences |
| | | Keeping up to date on developments in the field |
| | Personal organization and time management | Managing several tasks at once |
| | | Setting priorities |
| | | Allocating time efficiently to meet deadlines |
| | Personal strengths | Developing personal traits for dealing with day-to-day work situations. (For example, maintaining a high energy level, motivating oneself to function at an optimal level of performance, functioning in stressful situations, maintaining a positive attitude, being able to work independently, and responding appropriately to constructive criticism.) |
| | Problem-solving and analytical | Identifying, prioritizing, and solving problems, individually or in groups. It involves the ability to ask the right questions, sort out the many facets of a problem, and contribute ideas as well as answers regarding the problem. |
| Communicating | | |
| Interacting effectively with individuals and groups to facilitate the gathering, integrating, and conveying of information in many forms (example, verbal, written) | Interpersonal | Working well with others (superiors, subordinates, and peers), understanding their needs, and being sympathetic to them. |
| | Listening | Being attentive when others are speaking |
| | | Responding effectively to others comments during a conversation |
| | Oral communication | Effectively presenting information to others, either one-to-one or in groups |
| | Written communication | Effectively transferring information, either formally (through reports and business correspondence for example) or informally (through notes or memos). |

| | | |
|---|---|---|
| Managing people and tasks | | |
| Accomplishing the tasks at hand by planning, organizing, coordinating, and controlling both resources and people | Coordinating | Coordinating the work of peers and subordinates |
| | | Encouraging positive group relations |
| | Decision-making | Making timely decisions based on thorough assessments of the short- and long-term effects of decisions |
| | | Recognizing political and ethical implications of decisions |
| | | Identifying stakeholders and those who will be affected by the decisions made |
| | Leadership and influence | Directing and guiding others |
| | | Delegating work tasks to peers and subordinates in an effective manner |
| | | Motivating others to do their best |
| | Managing conflict | Identifying sources of conflict between oneself and others, and between other people |
| | | Taking steps to overcome disharmony |
| | Planning and organizing | Determining required tasks to meet objectives (strategic and tactical) |
| | | Assigning tasks to others appropriately |
| | | Monitoring progress made against the plan |
| | | Revising plans as necessary to include new information |
| Mobilizing innovation and change | | |
| Conceptualizing, as well as setting in motion, ways of initiating and managing change that involves significant departures from the current mode | Ability to conceptualize | Combining relevant information from a number of sources |
| | | Integrating information into more general frameworks |
| | | Applying information to new or broader contexts |
| | Creativity, innovation, change | Adapting to situations of change |
| | | Initiating change |
| | | Providing novel solutions to problems |
| | | Re-conceptualize roles in response to changing demands related to success |
| | Risk-taking – (taking reasonable job-related risks) | Recognizing alternative or different ways of meeting objectives |
| | | Recognizing potential negative outcomes |
| | | Monitoring progress toward the set objectives |
| | Visioning | Conceptualizing the future of the company |
| | | Providing innovative paths for the company to follow |

open-ended interviews and a survey it became clear that graduates and managers were much more concerned about the quality of generic skills such as written communication. Accordingly, the taxonomy in Table 12 was developed "to provide practitioners of higher education and workplace training with a common language of general skills needed by college and university graduates for life long learning and employability" (Evers et al. 1998, p. xviii). It concentrated on "generalist skills that higher education graduates need as a base supporting their specialist knowledge and skills" (Evers et al. 1998, p. xix).

## *Research Perspective: How Generic Graduate Attributes Are Understood*

This literature review began by looking at the full range of competencies desired in an engineering graduate. Its attention then moved increasingly toward the competencies needed for effective performance of work in general. The review will conclude by looking at an interesting Australian paper by Barrie (2006) which steps back from the concern to produce a list of graduate attributes and looks rather at what is understood by the term *generic graduate attributes* (GGA) – the so-called *soft skills*, *nontechnical competencies*, or *critical-cross-field outcomes*. This shift in focus is illuminating not only because the way generic attributes are understood affects how they are addressed in curricula, but also because it draws attention to the underlying nature of GGA and how they interrelate with the hard attributes of engineering knowledge and engineering application skills.

The paper by Barrie (2006) describes the findings of a phenomenographic study that was intended to identify the qualitatively different ways in which academics perceived the term *generic graduate attributes*. Four categories of perception were identified as follows:

1. GGA are *precursor skills* – "necessary basic … skills but irrelevant [to teaching in higher education] as they are a prerequisite for university entry" (p. 225). From this perspective, only disciplinary knowledge and skills should be included in the curriculum – they constitute the foreground – while GGA and other learning outcomes function merely as a *backdrop* and receive little formal attention in the tertiary classroom.
2. They are *complementary skills* – "useful skills that complement or round out disciplinary learning" (p. 226). In this perspective, GGA have a place in the curriculum but only as stand-alone modules that are not explicitly linked to disciplinary knowledge or skills.
3. They are *translation skills* – "abilities that let students translate, make, use, or apply disciplinary knowledge to the world" (p. 227). This acknowledges the role of GGA in the application of disciplinary knowledge and skills. Accordingly, their inclusion in the curriculum should, where appropriate, be explicitly linked to disciplinary knowledge.

4. They are *enabling skills* – "abilities that infuse and enable university learning and knowledge" (p. 229). Here the relation between GGA and disciplinary skills and knowledge is recognized to be more intimate to the extent that a graduate's level of competency is determined by the degree to which disciplinary skills and knowledge are interwoven and empowered by GGA.

These categories are of interest to this review in the following ways:

- They emphasize and clarify a number of points noted elsewhere in the review, especially in regard to the relative importance of competencies, bodies of knowledge, and technical and nontechnical knowledge and skills. As will be seen, they confirm perceptions that were important to but not clearly articulated in the development of the taxonomy.
- Barrie (2006) indicates that the progression from precursor to complimentary to translational to enabling skills suggests increasing recognition of the importance of generic attributes to the effectiveness of productive activity. In defining generic attributes as *precursor* or *complementary* the perception is that generic attributes are *discrete from* disciplinary knowledge. Defining them as *translational* and *enabling* means that they are perceived as *transformative of* disciplinary knowledge. For example, when generic attributes are defined as *translational* skills they are seen as *essential partners* of disciplinary knowledge in productive activity. When they are perceived as *enabling* skills, they are seen as the *primary and essential substrate* of productive activity that deploys and marshals disciplinary knowledge and skills in effective and appropriate ways.
- Interestingly, the perception of generic attributes as *precursor* skills makes the important point that the generic attributes that students bring with them to university are important and influential. As will be seen, this observation is a significant element in the motivation behind the development of the taxonomy.

## Part 3: The Taxonomy of Engineering Competencies

The taxonomy of engineering competencies was developed between 2001 and 2002 in the School of Chemical and Metallurgical Engineering at the University of the Witwatersrand, Johannesburg, South Africa (Woollacott 2003). It was formulated as part of a curriculum reform initiative set in the context of the major societal change emanating from the demise of apartheid and the considerable shift in the demographics and educational backgrounds of students entering higher education that was brought about by that change.

All the challenges associated with the massification of higher education experienced elsewhere in the world (Tinto 1975; Knight et al. 2003; Lomas 2004) are particularly acute in the South African educational landscape. In the references cited, the so-called *traditional student*[2] typically constitutes the minority of the student intake: in South Africa they constitute the majority (Woollacott et al. 2003). Levels of *under-preparedness* are high among incoming students as a result of socio-economic factors (Phurutse 2005) and the aftermath of apartheid education that had fostered an inferior education system for the majority of the population (Simpkins 2005). In addition, rates of attrition and academic failure were high and remain high (Pinto 2001; Letseka and Maile 2008).

As can be appreciated, the circumstances just described present significant challenges to any educational restructuring effort. The purpose of the taxonomy was to articulate needed graduate competencies in a way that was appropriate to the restructuring of the first-year program, particularly in regard to the introductory engineering course. How the taxonomy was developed and the rational behind its formulation is the subject of this part of the chapter.

## *The Issue of Responsiveness*

In Part 1, the four primary stakeholders in engineering education were identified based on the theory of curriculum responsiveness. To satisfy the requirement to be appropriately responsive to the interests of economic and disciplinary stakeholders, the taxonomy needed to embody the learning outcomes articulated in the national accreditation standards formulated by ECSA – the Engineering Council of South Africa (ECSA). (A shortened version of these has already been presented in Table 2.)

Given the context of a society deeply committed to the transformation of its citizenry, societal responsiveness was a particularly important issue. To satisfy the requirements to be appropriately responsive to societal needs, the *taxonomy* had to articulate competencies that had to do with personal transformation in terms of the issues articulated in ECSA standards and the issues raised in the section on responsiveness in Part 1.

Many of these issues have to do with the GGA addressed in the ECSA standards. However, these attributes articulate the end point of the educational process and give no attention to the diversity of student attributes at the start of that process. In addition, they do not stress sufficiently the competencies associated with "participating as responsible citizens" or of being an agent of social upliftment by virtue of being a competent graduate. The primary way the taxonomy addressed these concerns was to place particular emphasis on the engineer as a worker and as a leader.

[2]Ellsworth (1989, p. 297) in the context of higher education in the USA, refers to the mythical *traditional students* as "young, white, heterosexual, Christian, able-bodied, thin, middle-class, English-speaking, and male." To this description should be added the advantage of having received a good secondary education.

To satisfy the requirement of being appropriately responsive to learners, the taxonomy had to articulate graduate competencies in a way that took into account the diversity of the competencies of incoming students and how these needed to be developed in relation to required graduate attributes. To understand how the taxonomy addressed this concern, it is necessary to discuss the issue of learner responsiveness in the context of under-prepared students.

## *Quality and Responsiveness to the Learner When Under-Preparedness Is an Issue*

Engineering education facilitates a developmental journey that learners take to prepare themselves for a professional career. Each engineering program is designed according to assumptions about the competencies of the entrants to the program. There are formal expectations and informal ones. The formal assumptions are based on the specified outcomes of the relevant secondary education. The expectation is that the associated assessment procedures have been effective so that students who obtain the required qualifications actually posses the expected competencies. Informal expectations have to do with assumptions about proficiency in the language of instruction, study and life skills, and competencies "picked up" during secondary education, but not formally assessed. Examples of the latter include a good work ethic, reasonable questioning skills, and an inclination to learn by seeking understanding rather than by memorization.

Massification of education is usually accompanied by a diversification of the attributes of incoming students (Lomas 2004). Consequently, a mismatch frequently arises between the competencies of some of the incoming students and the assumed competencies on which existing educational programs are based. In a sense, the programs are under-prepared for the students (Masenya 1995). From the reverse point of view, incoming students may be under-prepared for the programs they enter in that their competencies are different to or compare negatively with the assumed competencies on which the curriculum is based (Masenya 1995; Woollacott et al. 2003).

At least some of the student attrition and academic failure among first year students can be shown to result from this mismatch rather than to other factors. This is demonstrated by the relative success of some of the educational interventions that have managed to improve the academic performance of under-prepared students (Hillman 1992; Pinto 2001; Knight et al. 2003).

A quality educational program will be appropriately responsive to the needs of its students. When under-preparedness is an issue, it suggests a need to restructure the program in such a way that it is better able to accommodate the diversity of the entering students. Such restructuring clearly should be based on a reevaluation of the academic, personal, and professional developmental journey the students must follow to achieve the desired learning outcomes and become competent engineering graduates.

Some of the elements of the developmental journey which under-prepared students must follow are easily identified and some are not. In some cases, *gaps* clearly exist in the knowledge and skills base of some students – for example, their proficiency in discipline knowledge and skills is inadequate (Rollnick et al. 1998; Taylor and Chou 1999; Malcolm and Zukas 2001; Mumba et al. 2002). In other cases, there is a lack of proficiency in the language of instruction (Miller et al. 1997; von Gruenewaldt 1999) and life-of-the-mind that is the focus of higher education. Restructuring here involves the provision of extra modules or support systems to address the gaps. This approach has been the primary tactic used in South Africa from 1980 onwards (Pinto 2001; Woollacott 2003; Woollacott 2006).

Many aspects of under-preparedness among students, however, are more subtle and are not manifested only in simple ways such as obvious gaps in knowledge and skills. In South Africa, for example, the learning practices of many incoming students have been deeply shaped by education approaches that emphasize and develop surface approaches to learning (Hillman 1992; Grayson 1996; Simelane 2006) – an emphasis on memorization, reliance on proficiency in "doing past papers," and the development of skill in recognizing patterns in exam questions and applying standardized solution methods (Simelane 2006). Students are strongly shaped by their past experiences. Years of immersion in schooling that promotes the development of such inappropriate learning practices leave a deep imprint that strongly affects how students view and engage with the world of tertiary learning. Such influences, combined with the impact of socio-economic disadvantage and, in extreme cases, limited exposure to the world of technology, result in student under-preparedness, the nature and impact of which is not easy to understand or to address effectively in educational restructuring.

How can a curriculum be appropriately responsive to learners who display the subtle features of under-preparedness just described? The primary motivation behind the development of the taxonomy was to address this question. The thinking that was involved will be explained in terms of GGA.

## Development of the Taxonomy

The motivation for developing the taxonomy was therefore to provide a better handle on what attributes needed to be developed, how they related to disciplinary knowledge and skills, what they might look like in embryonic form in incoming students, and how to be alert to inappropriate attributes. So as not to lose sight of the larger objectives of economic, disciplinary, and societal responsiveness, the taxonomy was developed as a statement pertaining to the full range of generic engineering competencies.

The strategy that seemed to offer the most effective way to achieve the objectives just outlined was to focus on the engineer as a worker – to focus on engineering work and the competencies and dispositions needed to do it well. In essence, the taxonomy was seen as a detailed answer to the broad question of what is involved in working as a competent engineer.

As noted above, the taxonomy was derived from a broad ranging literature review that included but looked beyond the sources that are normally accessed for the genesis of statements on graduate attributes. What is particularly significant about the *taxonomy* (Table 13) is that its organizing rationale is based on respected theory and its content is derived from both respected theory and strong research evidence.

In this regard, the following features of the taxonomy give weight to the claim that it is comprehensive in its coverage of the issues it addresses.

- The organization of its first level detail is based directly on a well-respected model of generic work (Campbell et al. 1993). That model claims to comprehensively describe the components of any type of job – a claim that has significant support in the field of industrial psychology and human resource management (Williams 2002, pp. 97–99). The nine items in the augmented Campbell model (Table 10) have been collapsed into five categories in the taxonomy. Organizing the taxonomy around these categories therefore provides a theory-supported claim that no aspect of work, at least at a generic level, has been overlooked.
- The content of the taxonomy is organized to give appropriate attention to three dimensions of competency – knowledge, skills, and dispositions. As indicated earlier, these correspond to the categories found in another Campbell model (Table 1) that claims to comprehensively describe the generic determinants of competency (Campbell et al. 1993).
- In the language of Barrie (2006), GGA are conceived primarily as enabling skills that are deeply embedded and interwoven with other attributes. Because the taxonomy is a classification of competencies, it makes distinctions that, to some extent, hide the interdependence between knowledge, skills, and dispositions.
- The descriptions of the knowledge and skills expected in a competent engineer are derived from the literature on accreditation standards and descriptions of engineering work as well as from published findings of surveys of stakeholder opinion.
- In the taxonomy, *dispositions* are used as a composite term that includes attitudes, traits, values, interests, orientations, commitments, and motivations. As the discussion about the generic elements of competency (Table 1) shows, it is a person's dispositions that determine the way in which that person's knowledge and skills are actually marshaled and brought to bear in the performance of his/her work.
- The seventh category in the taxonomy – advanced dispositions – was extracted from a competency model for technical professionals (Spencer and Spencer 1993, p. 163). As described earlier, the research on which the models were based was carefully structured to identify the characteristic behaviors that distinguished superior from ordinary performers. The reliability and comprehensiveness of these insights rests on the extensive range of the data collected and on the degree of rigor with which the data were analyzed and the research was conducted.

**Table 13** A taxonomy of engineering competencies (Woollacott 2003)

| 1st level categories | 2nd level categories | (3rd level categories and 4th level detail)<br>Ability, disposition or understanding |
|---|---|---|
| (A) Engineering – specific work | General engineering work | (1) Perform the different aspects of any engineering work or task namely initiating and planning the work/task, acquiring the resources needed, performing subtasks and evaluating and synthesizing results. |
| | | (2) Use appropriate engineering and computer methods, skills and tools and properly assess, analyze, and interpret the results they yield. |
| | | (3) Evaluate effectiveness, productivity, profitability, quality, service, impact or implications of any aspect of work done or planned and a disposition to do so. |
| | | (4) Arrange, sort, retrieve and properly assess data, knowledge, and ideas. |
| | Specialist engineering work | (5) Perform *analytical work* to solve existing and anticipated engineering problems and model relevant systems by (a) applying knowledge of mathematics and the natural, engineering and computational sciences, and (b) identifying, assessing, formulating, and solving convergent and divergent engineering problems in a creative and innovative way. |
| | | (6) Perform *design work* by converting concepts and information into detailed plans and specifications for the development, manufacture or operation of systems, processes, products or components that meet desired needs. |
| | | (7) Plan and perform *investigations* to (a) *test* that a design or product meets specifications, (b) *develop* products, components, systems or processes, or (c) *search for new knowledge* that can be applied for the advancement of engineering practice. |
| | Engineering mixed with other work | (8) Integrate specialist engineering work appropriately with work relating to core functions of an enterprise, management, administration, supervision, projects, sales, consulting, entrepreneurship or teaching in order to achieve the broader aims of the business enterprise or stated objectives in the different generic settings of engineering work. These are analysis, design, testing, development, maintenance, selling, research, line management, project management, consulting, teaching, and entrepreneurial endeavor. |
| (B) Non-engineering-specific work | General | (9) Perform tasks and execute behaviors not specific to one's particular job. |
| | | (10) Manage one's personal work effectively to ensure that all aspects are properly coordinated, are progressing in a satisfactory manner and that problems that arise are identified and are dealt with appropriately. |
| | | (11) Support and help peers and facilitate group functioning by being a good model, keeping the group directed and reinforcing participation by other group members. |
| | | (12) Ensure that the resources and capacity to do good work are maintained, sustained, and, where necessary, developed further. |

| | | |
|---|---|---|
| | Supervision, leadership | (13) Influence the performance of subordinates through interpersonal interaction and influence, modeling, goal-setting, coaching, and providing reinforcement. |
| | | (14) Function as a supervisor in the "line production" activities of the enterprise at the appropriate designated position in the supervision hierarchy. |
| | Management, administration | (15) Articulate goals for a unit/enterprise, organize people and resources to achieve these, monitor progress, help solve problems or overcome crises that stand in the way, control expenditures, represent the unit in dealing with other units or clients. |
| | | (16) Manage a project and ensure that it is completed successfully, on time and within budget. |
| (C) Communication | | (17) Effectively exchange, transmit, and express – verbally, graphically and in writing – knowledge and ideas to achieve set objectives when communicating with colleagues, peers, clients, superiors, subordinates, engineering audiences, or the larger community. |
| (D) Interpersonal interactions | Personal | (18) Interact effectively and positively with colleagues, clients, superiors, subordinates, engineering audiences and the larger community. |
| | Teamwork | (19) Function effectively on multidisciplinary teams through personal contributions and interactions with others that enhance their contributions. |
| (E) Dispositions | | Ability, disposition or understanding |
| (E1) Personal dispositions | General | (20) Agreeable personal style, characteristics, and self-management including maturity, initiative, enthusiasm, poise, appearance, values, goals, outlook, and motivation. |
| | | (21) Disposed to consistent commitment to all job tasks, to working at a high level of intensity and the willingness to keep working under adverse circumstances and to expend extra effort when required. |
| | | (22) Disposed to taking responsibility within own limits of competence. |
| | | (23) Interest and knowledge in contemporary issues. |
| | Discipline | (24) Disposed to maintaining personal disciplines and avoiding negative behaviors. |
| | | (25) Being critically aware of the need to act professionally and ethically. |
| | | (26) Being critically aware of the impact of engineering activity in a global/social setting. |
| (E2) Adaptive dispositions | Self- development | (27) Disposed to improving personal competencies in general. |
| | | (28) Understands nature/importance of effective learning skills and is able to apply them. |
| | | (29) Able to assess one's own performance effectively and accurately. |
| | | (30) Disposed to improving critical knowledge, skills, and dispositions in an effort to sustain or improve one's reputation and advancement prospects. |

(continued)

**Table 13** (continued)

| 1st level categories | 2nd level categories | (3rd level categories and 4th level detail) Ability, disposition or understanding |
|---|---|---|
| | Life-long learning | (31) Understands the requirement to maintain continued competence. |
| | | (32) Able to and disposed to engage in independent and interdependent life-long learning through well developed learning skills. |
| | Change management | (33) Able to manage the impact of change effectively and flexibly, and to engage in new learning in coping with change. |
| (E3) Advanced dispositions | Achievement orientation | (34) Works to meet required standards but also creates own measures of excellence. |
| Extracted from the competency dictionary and generalized competency model for "technical professionals" (Spencer and Spencer 1993, p. 163) | | (35) Disposed to improve performance or improve morale, revenues or customer satisfaction by making specific changes in the system or in own work methods. |
| | | (36) Sets and acts to reach challenging goals for self and others.[a] |
| | | (37) Innovates. |
| | Impact and influence | (38) Gives presentations tailored to audience, calculates the impact of own actions/words and adapts presentations or discussion to appeal to the interest and level of others. |
| | | (39) Shows concern with professional reputation. |
| | Conceptual thinking | (40) Recognizes key actions and underlying problems by observing discrepancies, trends and interrelationships, crucial differences, past discrepancies. |
| | | (41) Able to condense large amounts of information in a useful manner. |
| | | (42) Makes connections and patterns by pulling together ideas, issues, and observations into a single concept and identifies key issues in complex situations. |
| | Analytical thinking | (43) Anticipates obstacles, breaks problem apart systematically, makes logical conclusions, sees consequences and implications. |
| | Initiative | (44) Persists in problem solving when things do not go smoothly. Exceeds job description. Addresses problems before asked to. Creates opportunities. |
| | Self-confidence | (45) Expresses confidence in own judgment. Sees self as a causal agent, prime mover. |
| | | (46) Seeks challenges and independence, welcomes challenging assignments, seeks additional responsibility, states own position clearly and confidently. |
| | Interpersonal understanding | (47) Understands attitudes, interests, needs of others and is good at discerning the unspoken thoughts, concerns or feelings of others. |

| | |
|---|---|
| Concern for order | (48) Seeks clarity of roles and information, checks quality of work/information, keeps records and an organized workplace, monitors data, projects, and the work of others. |
| Information seeking | (49) Asks questions, personally investigates, digs deeper, calls or contacts others, does research, uses own ongoing systems, involves others. |
| Teamwork and cooperation | (50) Genuinely values others' input and expertise and is willing to learn from others. |
| | (51) Empowers others, encourages those who perform well and gives them credit. |
| Expertise | (52) Applies technical knowledge to achieve additional impact, goes beyond simply answering a question and helps resolve others' technical problems. |
| | (53) Exhibits active curiosity to discover new things, makes major efforts to acquire new skills and knowledge, and to maintain an extensive network of relevant contacts. |
| Service orientation | (54) Seeks information about the real, underlying needs of the client, beyond those expressed initially, and matches these to available (or customized) products or services. |

[a] "'Challenging' means there is a 50–50 chance of actually achieving the goal – it is a definite stretch, but not unrealistic or impossible" (Spencer and Spencer 1993, p. 26)

## Conclusions

The description of the development of the taxonomy has been presented as a case study that shows how a statement of graduate attributes has been formulated for a specific educational context. It has shown how that formulation has applied the principles of curriculum responsiveness as a basis for identifying the stakeholders of engineering education and how this basis has been pursued in the attempt to address the concerns of each stakeholder. It has shown that theory can be exploited to enhance the credibility of a statement about desired graduate attributes. It draws attention to the interrelatedness of the attributes that make up competency and make for productive activity.

Engineering practice is not static. Not only is new technology being developed all the time, but also there are shifts in emphasis, in the kinds of demands placed on engineers and, therefore, in how graduate engineers need to be educated. Consequently, the need from time to time to modify an existing curriculum or to develop a new one should be recognized to be a permanent feature of engineering education. Statements of the goals of engineering education which usually inform such educational restructuring should likewise be subjected to periodic review and updating. I trust that this case study and the literature review it embodies may serve as a useful resource for any involved in the future design, redesign, or delivery of engineering education programs.

## References

Accreditation Board for Engineering and Technology, Inc. (2007). *Criteria for accrediting engineering programs: Effective for evaluations during the 2008–2009 accreditation cycle.* Baltimore MD: ABET, Inc. Accessed January 2009, from http://www.abet.org/Linked%20Documents-UPDATE/Criteria%20and%20PP/E001%2008-09%20EAC%20Criteria%2012-04-07.pdf

American Society of Mechanical Engineers. (1995). *Mechanical engineering curriculum development initiative: Integrating the product realization process (PRP) into the undergraduate curriculum.* New York: American Society of Mechanical Engineers.

Anderson, L. W., Krathwohl, D. R., et al. (2001). *A taxonomy for learning, teaching, practices and change.* New York: Longman.

Argyris, C., & Schon, D. A. (1974). *Theory and practise.* San Francisco: Jossey-Bass.

Augustine, N. R. (1994). Socioengineering (and Augustine's second law thereof). *The Bridge*, Fall 1994, 3–14.

Bankel, J. K., Berggren, F., et al. (2003). The CDIO syllabus: A comparative study of expected student proficiency. *European Journal of Engineering Education, 28*, 297–317.

Barrie, S. C. (2006). Understanding what we mean by the generic attributes of graduates. *Higher Education, 51*, 215–241.

Benefield, L. D., Trentham, L. L., et al. (1997). Quality improvement in a college of engineering instructional program. *Journal of Engineering Education, 86*, 57–64.

Boeing. (1966). *Desired attributes of an engineer.* Accessed December 2008, from http://www.boeing.com/educationrelations/attributes.html

Busse, R. (1992). The new basics: Today's employers want the three R's and so much more. *Vocational Education Journal, 62*(5), 29–31.
Campbell, J. P., McCloy, R. A., Oppler, S. H., & Sager, C. E. (1993). A theory of performance. In N. Schmit & W. C. Borman (Eds.), *Personnel selection in organizations* (pp. 35–70). San Francisco: Jossey-Bass.
National Association of Colleges and Employers. (2008). *Job outlook 2009 survey*. Accessed December 2008, from http://www.naceweb.org/press/quick.htm#qualities
*Compact Oxford English Dictionary*. AskOsford.com. Accessed February 2009, from http://www.askoxford.com/concise_oed/competent?view=uk
Crawley, E. F. (2002). *Creating the CDIO Syllabus, a universal template for engineering education*. 32nd ASEE/IEEE Frontiers in Education Conference, Boston.
Crawley, E. F., Malmqvist, J., Östlund, S., & Brodeur, D. R. (2007). *Rethinking engineering education: The CDIO approach*. New York: Springer.
de Jager, H. G. & Nieuwenhuis, F. J. (2002). *The relation between the critical cross-field outcomes and the professional skills required by South African Technikon engineering graduates*. The 3rd South African Conference on Engineering Education, Durban, South Africa.
de Lange, G. (2000). *The identification of the most important non-technical skills required by entry-level engineering students when they assume employment*. The 2nd South African Conference on Engineering Education.
Department of Education (2007). *The Higher Education Qualifications Framework (HEQF): A single qualifications framework for a diverse system*. South African Department of Education.
Ellsworth, E. (1989). Why doesn't this feel empowering? Working through the repressive myths of critical pedagogy. *Harvard Educational Review, 59*(3), 297–324.
Engineering Council of South Africa (ECSA). (2004). *Whole qualification standard for Bachelor of Science in Engineering (BSc(Eng))/Bachelors of Engineering (BENG): NQF Level 7*. Authorized by Council, registered on the National Qualifications Framework: NLRD no 48694; Document: PE-61 Rev-2 26 July 2004. Accessed February 2009, from http://www.ecsa.co.za/documents/PE-61-r2.pdf
Evans, D. L., Beakley, G. C., et al. (1993). Attributes of engineering graduates and their impact on curriculum design. *Journal of Engineering Education, 82*, 203–211.
Evers, F. T., Rush, J. C., et al. (1998). *The bases of competence: Skills for lifelong learning and employability*. San Francisco: Jossey-Bass.
Grayson, D. J. (1996). A holistic approach to preparing disadvantaged students to succeed in tertiary science studies. Part 1: Design of the Science Foundation Programme (SFP). *International Journal of Science Education, 18*(8), 993–1013.
Grimson, J. (2002). Re-engineering the curriculum in the 21st century. *European Journal of Education, 27*(1), 31–37.
Hesketh, B., & Neal, A. (1999). Technology and performance. In D. R. Ilgen & E. D. Pulakos (Eds.), *The changing nature of performance*. Hoboken, NJ: Wiley.
Hillman, J. C. (1992). *The Wits Pre-University Bursary Scheme (PBS): A bridging year for disadvantaged engineering students*. IEEE 3rd Conference in Africa (Africon), Swaziland.
International Engineering Alliance. (2005). Graduate attributes and professional competencies, Ver 1.1, 13 June 2005.
Kemp, N. D. (1999). The identification of the most important non-technical skills required by entry-level engineering students when they assume employment. *South African Journal of Higher Education/SATHO, 13*(1), 178–186.
Knight, D. W., Carlson, L. E. et al. (2003). Staying in engineering: Impact of a hands-on, team-based, first-year projects course on student retention. *Proceedings of the 2003 American Society for Engineering Education Annual Conference*. American Society for Engineering Education.
Koen, P. A. & Kohli, P. (1998). *ABET2000: What are the most important criteria to the supervisors of new engineering graduates*. 1998 ASEE Annual Conference and Exposition, Seattle, WA.
Landis, R. B. (2007). *Studying engineering: A road map to a rewarding career*. Los Angeles: Legal Books Distributing.

Lang, J. D., Cruse, S., et al. (1999). Industry expectations of new engineers: A survey to assist curriculum designers. *Journal of Engineering Education, 88*, 43–51.
Lattuca, I. R., Strauss, L. C., et al. (2006). Getting in sync: Faculty and employer perceptions from the national study of EC2000. *International Journal of Engineering Education, 22*, 460–469.
Letseka, M. & Maile, S. (2008). *High drop-out rates: A threat to South Africa's Future. HSRC Policy Brief.* Human Sciences Research Council of South Africa.
Lomas, L. (2004). Embedding quality: The challenges for higher education. *Quality Assurance in Education, 12*(4), 157–165.
London, M., & Mone, E. M. (1999). Continuous learning. In D. R. Ilgen & E. D. Pulakos (Eds.), *The changing nature of performance*. Hoboken, NJ: Wiley.
Maillardet, F. (2004). What outcome is engineering education trying to achieve? In C. Baillie & I. Moore (Eds.), *Effective learning and teaching in engineering*. London: Taylor & Francis Group.
Malcolm, J., & Zukas, M. (2001). Bridging pedagogical gaps: Conceptual discontinuities in higher education. *Teaching in Higher Education, 6*(1), 34–42. doi:10.1080/13562510020029581.
Masenya, D. I. (1995). Reconceptualizing the academic discourse: Underprepared students or institutions or both? *Academic Development, 1*(2), 99–105.
Miller, R., Bradbury, J., et al. (1997). Academic performance of first and second language students: Kinds of assessments. *South African Journal of Higher Education/SATHO, 11*(2), 70–79.
Moll, I. (2004). Curriculum responsiveness: The anatomy of a concept. In H. Griesel (Ed.), *Curriculum responsiveness: Case studies in higher education* (pp. 1–19). Pretoria: SAUVCA, South African Vice-Chancellors Association.
Mumba, F. K., Rollnick, M., et al. (2002). How wide is the gap between high school and first year chemistry at the University of the Witwatersrand. *South African Journal of Higher Education, 16*(3), 148–157.
National Society of Professional Engineers. (1992). *Engineering education issues: Report on surveys of opinions by engineering deans and employers of engineering graduates on the first professional degree*. National Society of Professional Engineers (NSPE) Publication No. 3059. Alexandria, VA.
Natriello, G. (1989). *What do employers want in entry-level workers: An assessment of the evidence*. New York: Columbia University Press.
Passow, H. J. (2007). *What competencies should engineering programmes emphasize? A meta-analysis of practitioners' opinions informs curricula design*. 3rd International CDIO Conference. Cambridge, MA: MIT.
Phurutse, M. C. (2005). *Factors affecting teaching and learning in South African public schools*. Cape Town: University of Cape Town.
Pinto, D. (2001). (Ed.). *Directory of science, engineering and technology foundation programs*. Wits, South Africa: College of Science, University of the Witwatersrand Johannesburg, Central Printing Unit.
Rollnick, M., Manyatsi, S., et al. (1998). A model for studying gaps in education: A Swaziland case study in the learning of chemistry. *International Journal of Educational Development, 18*(6), 453–465.
Saunders-Smith, G. N. (2005). *The secret of their success: What factors determine the career success of an aerospace engineer trained in the Netherlands*. ASEE Annual Conference and Exposition. Portland: ASEE.
Shea, J. E. (1997). An integrated approach to engineering curricula improvement with multi-objective decision modelling and linear programming (Doctoral thesis). *Dissertation Abstracts International* A58:1649.
Simelane, Z. F. (2006). *Identification and classification of incoming learning behaviours amongst a sample of first year, English second language, engineering students: A case study*. Wits, South Africa: University of the Witwatersrand, Johannesburg.
Simpkins, C. E. W. (2005). *Learner performance in South Africa: Social and economic determinants of success in language and mathematics*. Cape Town South Africa: HSRC Press.
Sinha, M. N., & Willborn, W. O. (1985). *The management of quality assurance*. New York: Wiley.

Skakoon, J. G., & King, W. J. (2001). *The unwritten laws of engineering: Revised and updated.* New York: ASME Press.
South African Qualifications Authority. (1997). *SAQA Bulletin.* Department of Education, South African Qualifications Authority.
South African Qualifications Authority. (2000). National qualification framework and curriculum development. Accessed December 2008, from http://www.saqa.org.za/structure/nqf/docs/curriculum_dev.pdf
Spencer, L. M., & Spencer, S. M. (1993). *Competence at work: Models of superior performance.* New York: Wiley.
Taylor, D. W. & T. F. Chou (1999). *Evidence of the gap between student's learning approaches and instructors' teaching approaches in accounting education.* 8th Annual Teaching Learning Forum: Teaching in the disciplines/learning in context. The University of Western Australia.
Tinto, V. (1975). Drop-out from higher education. A theoretical synthesis of recent data. *Review of Educational Research, 45*(1), 89–125.
Turley, R. T. (1992). Essential competencies of exceptional professional software engineers (Doctoral thesis). *Dissertation Abstracts International* B53:400.
von Gruenewaldt, J. T. (1999). Achieving academic literacy in a second language: South Africa's educational predicament. *South African Journal of Higher Education/SATHO, 13*(1), 205–212.
Williams, R. S. (2002). *Managing employee performance: Design and implementation in organisations.* Cengage Learning EMEA.
Woollacott, L. (2003). *Dealing with under-preparedness in engineering education. Part 1: Defining the goal: A taxonomy of engineering competency.* Originally in WFEO/ASEE e-Conference, from http://web.wits.ac.za/library/electronicthesesdissertations.html.
Woollacott, L. (2006). *From academic development to student development: A scenario for a transformed curriculum in South African engineering educationation.* 3rd African Regional Conference on Engineering Education, Pretoria, South Africa.
Woollacott, L. (2007). *The goals of engineering education: A rationale for a universal document based on the CDIO syllabus and the taxonomy of engineering competencies.* 3rd International CDIO Conference, Boston.
Woollacott, L., Henning, L. et al. (2003). *Addressing under-preparedness in entrants to the chemical engineering program at Wits.* South African Chemical Engineering Congress, Sun City, South Africa.
World Chemical Engineering Council. (2004). *How does chemical engineering education meet the requirements of employment?* Short Report. World Chemical Engineering Council Secretariat. Accessed December 2008, from http://www.chemengcouncil.org/chemengworld_media/Downloads/short_report.pdf
Young, J. L. (1986). What competencies do employers really need: A review of three studies. *Journal of Career Development, 12*(3), 240–244.

## Professional Engineering Bodies: Web Sites for Statements on Accreditation Standards

Accreditation Board for Engineering and Technology, United States, http://www.abet.org
Canadian Council of Professional Engineers, http://www.engineerscanada.ca
Engineering Council of South Africa, http://www.ecsa.co.za
Engineering Council UK, United Kingdom, http://www.engc.org.uk
Engineers Australia, http://www.engineersaustralia.org.au
Institution of Professional Engineers, New Zealand, http://www.ipenz.org.nz
International Engineering Agreements: http://www.ieagreements.com

# Future Direction

# Internal and External Quality Assurance Approaches for Improvement and Accountability: A Conceptual Framework

**Peter J. Gray and Arun Patil**

**Abstract** The keys to advancing Quality Assurance are to, first, strike a balance between Internal/External and Improvement/Accountability emphases; second, recognize the value of various Quality Assurance approaches, for different purposes; and, third, acknowledge the trade-offs and tensions inherent in various approaches. The changes implied by Quality Assurance must start at the local level, i.e., individual courses and programs of study such as Engineering Education. The task of documenting such changes and, thereby, recognizing the impact of Quality Assurance policies and practices (Accreditation and Evaluation or Assessment) makes it necessary to use different metrics at different levels of a higher education institution. This is where the value of the conceptual framework described in this chapter can be seen, since it acknowledges a wide range of Quality Assurance approaches, thus providing a means of engaging all stakeholders in a constructive way about how to define quality, how to document it, and how to make needed improvements.

## Introduction

It is especially important to appreciate the range of Quality Assurance approaches that are available in higher education because recent developments have focused the discussion on the extreme ends of the two continua (internal vs. external and improvement vs. accountability). As noted in the chapter "The Background of Quality Assurance in Higher Education and Engineering Education" by Gray, Patil, and Codner, the most urgent current demands focus on the information on college students' performance that is publicly available and comparable across institutions and engineering programs (*external*), used to inform policy and resource allocation decisions, and used to inform consumer choice (*accountability*). In addition, to best

P.J. Gray (✉)
Faculty Enhancement Center, United States Naval Academy, 589 McNair Road, 10M, Annapolis, MD 21402, USA
e-mail: pgray@usna.edu

A.S. Patil and P.J. Gray (eds.), *Engineering Education Quality Assurance: A Global Perspective*,
DOI 10.1007/978-1-4419-0555-0_22, 

meet the twin demands of *improvement* and *accountability*, it is essential to be clear about the language and power arrangements implied by the various points along the two continua and, thereby understand the value, one might even say the necessity, of including all of the variations in the operational definition of Quality Assurance.

Figure 1 and the following discussion are intended to provide a conceptual framework for organizing, understanding, and harmonizing the range of "systematic management and assessment procedures adopted by a higher education institution or system to monitor performance and to ensure achievement of quality outputs or improved quality" (Harman and Meek 2000, p. 4).

## The Two Continua

There are two sets of concepts related to Quality Assurance. One describes the *ownership* of the process, from stakeholders *internal* to those *external* to an institution, degree type, or program. The second continuum describes the *purpose* of Quality Assurance ranging from *improvement* to *accountability*. The reason that these are continua and not discrete categories is that there are shades of gray from one end to the other. Recognizing and appreciating the sometimes subtle differences and interplay among the approaches that can be described in relation to these two continua are important to advancing Quality Assurance.

See Ewell (1991) and Terenzini (1989) for earlier examples of taxonomies for conceptualizing and describing assessment activities in higher education.

At one end of the horizontal Internal/External continuum in Fig. 1 are those Quality Assurance approaches that are under the control of and are designed to serve the higher education community internal to an institution, degree type, or program. This end of the continuum signifies an *internal locus of control*. At the other end of the continuum are approaches to monitor performance and to ensure

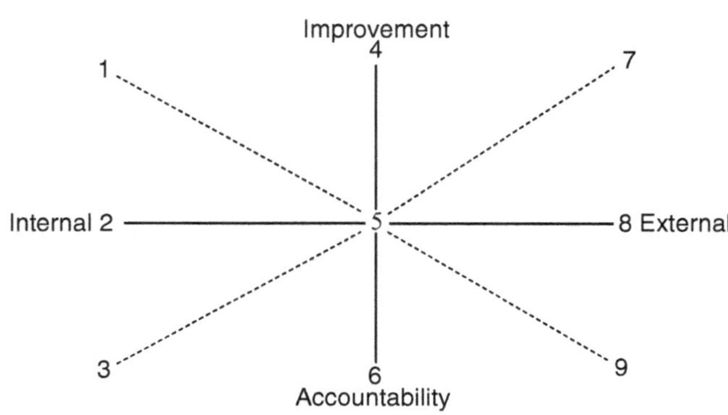


**Fig. 1** A conceptual framework for describing approaches to Quality Assurance in higher education

achievement of quality outputs or improved quality organized by those outside of higher education. This suggests an *external locus of control.*

The vertical axis in the figure is an Improvement/Accountability continuum. The top is suggests information intended to guide improvement and at the bottom is information that documents institutional, degree-type, or program compliance with implicit or explicit criteria and standards, or accountability.

## *Points along the Continua*

While fixed points are used to provide illustrative examples, there really is a range of Quality Assurance approaches along the Internal/External and Improvement/Accountability continua (solid lines) and among the points (dotted lines). The intent of the following examples is to illustrate how the framework can be used to describe and, therefore, better understand and appreciate the universe of approaches to Quality Assurance.

*Point 1* represents local and often informal initiatives conducted by those within an institution, degree type, or program, regularly or on an ad hoc basis as the need arises, that are solely intended to produce information to be used by internal audiences for developing or improving quality. This includes classroom assessment conducted by individual instructors for the purpose of improving teaching and learning and the ongoing evaluation of a program by those directly responsible for its operation.

*Point 2* represents more formal and perhaps regular initiatives that while controlled entirely locally are intended to strike a balance between improvement and accountability. As an internally organized approach, the results (suggestions for improvement and any recommendations regarding continuation) would go to local administrators who have oversight responsibilities. For example, a course initiation and review process may be part of an internal institutional monitoring scheme with certain expectations and responsibilities. These might include criteria and standards that form the basis for approval of new or revised courses by an institution-wide curriculum committee.

*Point 3* represents local approaches that are entirely intended to monitor activities without an explicit intention to provide feedback for improvement. Such an approach might involve the production of an annual status report, based on fixed categories, intended only to document compliance with reporting requirements. In some cases, such reports may be used to determine internal allocation of resources based on the extent or value of the reported information. For example, an individual faculty member's annual report of student course evaluation results, publications, and service activities may be tied to merit pay or salary increases and, at the program level, the annual reporting of accumulated accomplishments of its faculty may influence internal budget allocations.

*Point 4* indicates an improvement exercise that is conducted using a combination of internal and external standards and criteria. For example, in revising an institution's mission and goals or a program curriculum, a local committee might use standards and criteria developed by external organizations such as associations of similar colleges and universities or discipline societies. Another example is a program review that is a periodic internal requirement. Such a review may be organized around a set of externally *normed* criteria that a team of external experts uses to examine the program. The primary purpose of the review is to provide guidance for the faculty members to improve the program without any implications for *accountability* or the continued *right to exist.*

*Point 5* is the typical Quality Assurance approach for the purpose of institutional or disciplinary accreditation. A self-study is conducted internally based on the established external standards and criteria and then an external team of experts reviews the self-study, visits the campus, and provides both suggestions for improvement and recommendations regarding accreditation status. As an externally driven process based on the policies and procedures of the accrediting body, suggestions for *improvement* are only given to *internal* audiences and recommendations regarding *accreditation* are made to an *external* agency, thus holding the institution, degree type, or program *accountable* for meeting the standards. Details of any suggestions for improvement are not made *public*, but the accreditation status may be, for example, fully or unconditionally accredited, provisionally accredited with reservations, or not accredited. Of course, these categories have implications regarding the urgency of improvement efforts embedded in the suggestions by the visiting team.

*Point 6* is similar to Point 3 in that it is intended to monitor activities without any explicit intention of providing feedback for improvement. However, as opposed to the entirely *internal* approach in Point 3, the focus of this example is prescribed by *external* standards and criteria. Point 6 may involve the *internal* production of an annual report that is intended only to document compliance with *external* reporting requirements. The report goes to both *internal* and *external* audiences. The internal audience may include program and central administrators responsible for ensuring compliance and external audiences responsible for monitoring compliance, for example a Ministry of Education or a specific governmental or nongovernmental agency with oversight responsibility for a particular area such as Engineering Education.

*Point 7* suggests an externally organized review intended only for improvement. It might be prompted by information coming from a variety of sources, perhaps even an accreditation review, and is conducted by an external agency in order to guide needed changes in an institution, degree type, or program. For example, sometimes an accreditation visit results in a suggestion that provides direction for a particular improvement to be made within a specified period of time. The results of the action are then reviewed in light of the improvement called for on the timetable specified.

*Point 8* might involve an evaluation conducted by an external agency with no internal involvement such as a self-study. This sort of inspection may result in both suggestions for improvement to internal audiences and the dissemination of information to external audiences for accountability purposes. Point 8 may be viewed as an approach similar to a financial audit where budgeting and accounting procedures are examined for compliance with Generally Accepted Accounting Principles (GAAP). These are a set of rules used to standardize the reporting of financial statements and to give quality ratings of organizations. Feedback is provided regarding necessary improvement and recommendations are made as to the quality of the budgeting and accounting procedures.

*Point 9* is epitomized by the rankings and league tables that are developed by external agencies such as *US News and World Report* or the *Times Higher Education Supplement*. These assessments are conducted in order to compare and contrast institutions, degree types, or programs on a set of externally designated criteria using a process that is entirely under the control of external agencies. There is little if any influence by internal stakeholders on the criteria used and, as a consequence, the relevance of the results for improvement is tenuous at best.

The value of the conceptual framework in Fig. 1 is that it provides examples of different approaches to Quality Assurance that can be taken, thus giving institutions and programs much needed flexibility to meet the competing *internal* and *external* demands for *improvement* and *accountability*.

## *Harmonizing QA Approaches*

The general idea behind the conceptual framework is that different methods (*means*) should be used to provide information that is needed for different *ends*, i.e., to support *Improvement* and to comply with *Accountability* demands. For example, the information needed to facilitate most local Quality Assurance for Improvement will be idiosyncratic to a course or program. Evaluations or assessments should be embedded in the regular processes (e.g., exams, assignments, and projects) and address local questions about Quality Assurance (i.e., be *authentic*), such as, *are students learning what we expect them to learn from a given course or program of study*?

While the detailed results of such efforts may not be easily summarized across courses, programs, or institutions, it is possible to develop a matrix of outcomes and the educational activities that are intended to foster them. This exercise can be repeated from the course to the program level (e.g., Engineering Education major) and on to institutional level so that, in the end very general information about the accomplishment of a common mission and goals can be reported that is based on substantial data at the most local level. In this way, locally relevant data developed for the purpose of improvement and under the control of instructors can be reexamined as the need arises to provide summarized *information* to internal and external audiences related to accountability issues.

Similarly, even an external Accountability measure such as a ranking or league table, if it can be linked to institutional and program goals, can provide general information as to where strengths and weaknesses lie and, therefore, guide further investigations to determine the extent, causes, and possible responses needed to improve quality. For example, standard instruments like the National Survey of Student Engagement in the USA or the National Student Survey in the UK, and the Course Experience Questionnaires in Australia may identify issues such as *time spent preparing for class* or the *effectiveness of feedback to students* as weaknesses in comparison with other comparable institutions or programs. While such information is not specific enough to suggest particular improvements, it can be used to stimulate a discussion about these topics and if verified by local information can guide appropriate changes.

Of course, caution is warranted in the relation to such approaches. They are ostensibly intended to hold higher education accountable to the public and to justify confidence and may or may not provide any insights for improvement. They may, in fact, inhibit the improvement of teaching and learning (Hoecht 2006) as well as the accomplishment of other goals such as the inclusion of underrepresented groups in higher education (Clarke 2007). Nevertheless, the results should be examined in relation to institutional and/or program goals in order to provide a context for their interpretation.

## Putting the Conceptual Framework to Use

At the middle of the framework is the typical accreditation approach to Quality Assurance. It provides an appropriate model for the integration of all the other approaches suggested by the illustrative points. As described by the European Federation of National Engineering Associations (FEANI) (ESOEPE 2005):

- Accreditation is the primary Quality Assurance process used to ensure the suitability of an educational program as the entry route to the engineering profession.
- Accreditation involves a periodic audit against published standards of the engineering education provided by a particular course or program.
- It is essentially a peer review process, undertaken by appropriately trained and independent panels comprising both engineering teachers and engineers from industry.
- The process normally involves both scrutiny of data and a structured visit to the educational institution.

### *Mission and Goals*

Point 5 implies a *harmonizing* of Quality Assurance approaches, i.e., evaluation and assessment methods from the course to the institutional level for internal and external purposes of improvement and accountability. *Harmonizing*, rather than *standardizing*, suggests that the same methods, techniques, and instruments do not have to be used in all circumstances, but there should be a synergy of approaches that results in the gathering of a body of information related to institutional or program effectiveness in meeting its goals and fulfilling its mission.

In many accreditation schemes, synergy is accomplished by having, first, a common focus such as the mission and goals of an institution, degree type or program; second, a structure for harmonizing the evaluation and assessment methods used to gather, analyze, and report specific results; and, third, a process for summarizing findings across efforts in relation to the mission and goals.

Therefore, the first step in creating a Quality Assurance structure is to reach consensus on institutional and/or program mission and goals. The second step is to conduct an audit of existing QA approaches.

## *QA Audit*

Inevitably there are many different Quality Assurance approaches underway at any given time in an institution or program. The framework provides a way to organize them.

Undertaking a QA audit can provide a clear picture of what is currently occurring. Starting where you are and recognizing previous work and its success not only acknowledges all of the hard work that people have done related to Quality Assurance, but also shows that the institution or program values quality and can make changes in teaching and learning as well as other areas in order to enhance the accomplishment of its mission and goals.

By using the framework in Fig. 1 as an organizing structure, some approaches that were not considered to be part of the Quality Assurance process may be identified along with gaps that can be filled. As a result, a comprehensive set of Quality Assurance tools will be available to address internal and external calls for improvement and accountability.

## *Adoption of an Innovation: Planned Change and Leadership*

Quality Assurance, especially as suggested by the whole range of approaches in the framework, is new to many people in higher education. And, to the extent that this comprehensive conceptualization is different from traditional approaches to Quality Assurance, it can be viewed as an innovation with resulting resistance if not outright hostility. By focusing on the many ways that an institution or program has enhanced quality in the past, it is possible to productively engage people in determining how to “continue to change and grow in order to adapt to current conditions” (Gray 1997, p. 5).

However, acknowledging the range of approaches embodied in the framework suggests a cultural change in higher education. Therefore, sustained and sensitive leadership is needed to guide a process of planned change for adopting this innovation.

As noted by Curry (Gray 1997), leaders can facilitate change by providing a conducive climate, helping to define and shape issues through translating them in a way consistent with the local culture, pointing out the success of past efforts, building community-wide coalitions in support of change, providing funding and other incentives for participating in the change process, being a sponsor and facilitator of the change but sharing leadership with others, and being a visionary and helping others see the positive effects of the change.

Educators are by nature concerned with quality. By emphasizing how engagement in a wide range of Quality Assurance approaches can foster continual improvement of teaching and learning, a leader can help faculty members understand that this is consistent with their own interests and a legitimate part of their professional role and responsibility.

See the chapter "Using Soft Systems Thinking to Confront the Politics of Innovation in Engineering Education" by Eijkman, Kayali, and Yeomans for a thorough discussion of how to confront the politics of innovation in engineering education.

## *Using Accreditation as the Focal Point*

Because accreditation is the most common experience of formal Quality Assurance in Engineering Education and in many higher education institutions, it is possible to use it to draw together all of the various Quality Assurance approaches into a coordinated whole. Its place at the center of the framework implies that it can be used as a focal point for Quality Assurance.

Accreditation can be used to build a structure where internal improvement efforts are documented and summarized, and their impact shared, and where external accountability information can be interpreted and used as is appropriate. As a periodic audit against published standards, it provides the structure for organizing all of the formal and informal Quality Assurance approaches suggested by the framework in Fig. 1 to provide information that forms a complete narrative of a program. The themes of this narrative are the demonstration of the extent that the mission and goals are being achieved, based on evidence gathered through evaluation or assessment, and the description of the resulting actions that are being taken to foster continuous improvement.

The typical process that underlies Assessment is shown in Fig. 2. Based on the missions and goals, various institutional practices involving teaching and learning as well as administrative functions and facilities are evaluated in order to provide feedback for improvement. The Agencies of Assessment are the internal and external stakeholders who are responsible for implementing the assessment processes concerning Quality Assurance issues and questions. A central steering committee within a program and/or institution usually provides the overarching harmonization of the various assessments and coordinates the synthesis and reporting of results.

There are two categories of benefits of accreditation: academic (instructors and students) and administrative (programs and institutional). The potential benefits of accreditation for instructors and students may include:

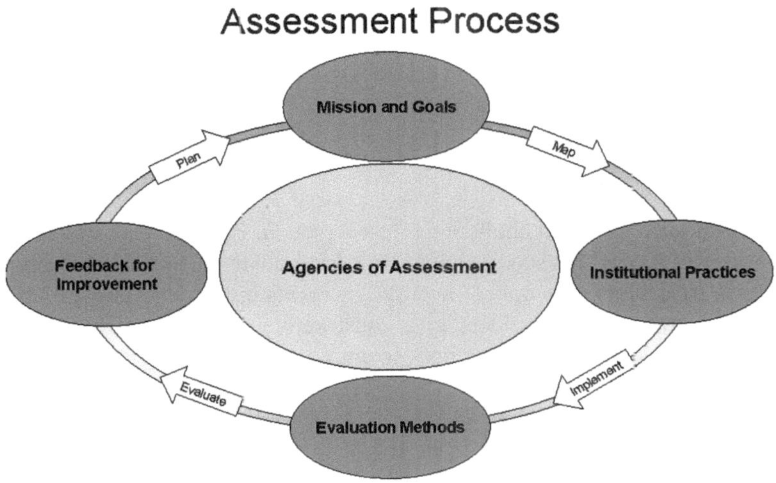


**Fig. 2** Assessment process (Based on the USNA Faculty Assessment Committee model)

- Better design and implementation of course curriculum and educational programs
- Measurement of learning outcomes of students
- Identification of strengths and weaknesses in teaching/learning processes as well as classroom and laboratory facilities
- Information to promote external interactions and the placement of students
- Identification of opportunities for the professional development of instructors and students

The potential benefits of accreditation for programs and institutions may include:

- Documentation of the accomplishment of missions and goals
- Guidance for the development and enhancement of instructional resources
- Identification of reliable information for use with internal and external audiences
- Opportunities for national and international networking
- Improvement of institutional reputation and prestige in the global environment

Focusing on these benefits will help to provide the rationale for *institutionalizing* a process like that shown in Fig. 2. This in turn would make the preparation for each accreditation review much easier. A permanent Quality Assurance structure may, for example, take the form of a single QA coordinator and/or a group of *outcomes champions* for the various institutional or programmatic goals. Or it may involve creating an Office of Institutional Effectiveness with responsibility for harmonizing all of the approaches to Quality Assurance across an institution.

Once a permanent structure is established, it is a matter of taking a snapshot of an ongoing QA process each time that an accreditation review takes place or other Quality Assurance questions are raised. In addition, the wealth of information, the depth of interpretation, and the long-term documentation of the impact of changes that all come from the consistent implementation of a variety of Quality Assurance approaches are, ultimately, the best way to guide *internal improvement* and satisfy *external* demands for *accountability*.

## Conclusion

Quality Assurance's biggest challenge is to balance the competing assumptions and expectations of its various stakeholders. By taking advantage of all of the possible approaches to Quality Assurance, a program or institution can engage all of its *internal* and *external* stakeholders in a constructive dialog about how to define quality, how to document it, and how to make needed *improvements* and meet demands for *accountability*.

## References

Clarke, M. (2007). The impact of higher education rankings on student access, Choice, and Opportunity. In *College and university ranking systems: Global perspectives and American challenges* (pp. 35–48). Washington, DC: Institute for Higher Education Policy.

Ewell, P. T. (1991). To capture the ineffable: New forms of assessment in higher education. *Review of Research in Education*, . doi:10.3102/0091732X017001075.

Gray, P. J. (1997). Viewing assessment as an innovation: Leadership and the change process. In P. J. Gray & T. W. Banta (Eds.), *The campus-level impact of assessment: Progress, problems, and possibilities.* (New directions for higher education, Vol. 100, pp. 5–15). San Fransisco, CA: Jossey-Bass.

Harman, G. S., & Meek, V. L. (2000). *Repositioning quality assurance and accreditation in Australian higher education.* Canberra Commonwealth of Australia: Department of Education Training and Youth Affairs (DETYA). Evaluations and Investigations Programme Higher Education Division, AusInfo.

Hoecht, A. (2006). Quality assurance in UK higher education: Issues of trust, control, professional autonomy and accountability. *Higher Education*, . doi:10.1007/s10734-004-2533-2.

Terenzini, P. T. (1989). Assessment with open eyes. *Journal of Higher Education, 60*(November/December), 644–664.

# Index

## H



## I



## L



## M

FSC
www.fsc.org
MIX
Papier | Fördert
gute Waldnutzung
FSC® C083411

Kolibri 360 GmbH
Ferdinand-Jühlke-Straße 7
99095 Erfurt, Deutschland
produktsicherheit@kolibri360.de